Chavan–Heisenberg Ripple Uncertainty Principle

Sandeep Chavan

Published by SANDEEP CHAVAN, 2025.

CHAVAN–HEISENBERG RIPPLE UNCERTAINTY PRINCIPLE

First edition. July 5, 2025.

ISBN: 979-8230021728

Written by Sandeep Chavan.

Table of Contents

To **Werner Heisenberg,**with immense respect and a ripple of humor.

Because if you hadn't arrived on time,they might've built a bungalow for every particle—complete with a permanent address, a driveway, and a zip code.

Your "uncertainty" gave physics a pause,and that pause gave us time to listen deeper.

Thank you for keeping science... wonderfully unsettled.

— Sandeep Chavan

"What we called randomness was just ripple protocol—misread, mistranslated, and rushed."— Chavan's Law

Disclaimer

This book presents an original theoretical framework—**Universal Energy Dynamics (UED)**—that reinterprets classical and quantum phenomena through the lens of ripple alignment, perception thresholds, and field coherence. It challenges widely accepted scientific conventions, including interpretations of the Heisenberg Uncertainty Principle, wave-particle duality, and quantum randomness.

The author does not claim that this work replaces established science, nor does it assert experimental proof of every concept discussed herein. Instead, it offers a **conceptual, philosophical, and mathematical alternative** that seeks to inspire inquiry, encourage critical thinking, and open the door to future validations.

This work is **independent research**. All views expressed are those of the author and may not align with mainstream academic institutions or scientific organizations. While care has been taken to remain rigorous and logically consistent, **readers are urged to evaluate the claims within the proper context**—as an emerging paradigm still under exploration and refinement.

No part of this book is intended to diagnose, treat, or resolve any scientific, psychological, or personal issue. The metaphors used—such as ripple fields, misalignment, or perception thresholds—are illustrative tools designed to assist understanding, not to prescribe behavior or replace professional judgment.

Any resemblance to existing theories, academic debates, or named individuals—living or deceased—is either intentional for comparative clarity or purely coincidental.

In short: This is not the end of science. It is an invitation to expand it.

Opening Note

This is not a book written to impress. It is written to **express**—an understanding born not from institutional laboratories, but from independent ripples of thought, years of teaching, deep observation, and fearless questioning of the obvious.

In the world of modern physics, mystery often hides behind elegant math and poetic uncertainty. Yet we continue to ask—why do we see what we see? Why do we miss what seems hidden? And why, even with all our knowledge, do we still mistake **alignment** for **luck**, **delay** for **randomness**, and **clarity** for **certainty**?

This work is not a rejection of Heisenberg or quantum physics. It is an expansion—a ripple forward. Where uncertainty was once treated as a fundamental wall, we now offer a doorway: one that opens not with more force or faster tools, but with better **fit**.

Here, you will meet **RAC** and **ATT**—two ideas that don't overthrow old science, but realign it. You'll learn why precision fails when the ripple hasn't arrived, and why consciousness itself is a function of **when**, not just **what**.

If you're tired of paradoxes that ask you to suspend reason, if you feel that reality shouldn't contradict itself just to stay interesting—this book is for you.

We are not theorizing from above. We are observing from within. This is not a call to believe. It is an invitation to perceive—with **patience**, **alignment**, and a science that finally listens.

— *Sandeep*

Preface

Chavan–Heisenberg Ripple Uncertainty Principle

For over a century, the Uncertainty Principle has stood like a boundary stone—defining the edge of what we could measure, predict, or even believe to be real. Heisenberg declared that precision had a price, that the act of observing fractured reality itself. But what if this "fracture" was not a break in nature, but in understanding?

This book introduces not just a correction, but a completion.

The **Chavan–Heisenberg Ripple Uncertainty Principle** does not discard the quantum view. It reveals the ripple behind it. It dares to suggest that uncertainty is not a built-in blur, but a signal still aligning. That misreading a ripple is not the same as collapsing a wave.

Here, we offer not just another framework—but a shift in perception. Reality doesn't hide. It waits. Consequence doesn't jump—it resolves. And the observer doesn't disrupt—it fits—or doesn't.

This principle rests on two revolutionary anchors:

- **RAC** – the Ripple Alignment Condition
- **ATT** – the Alignment Threshold Time

Together, they form a living model of perception, action, and consequence. Not abstract probabilities, but ripple-based readiness. You don't see because something exists—you see because it fits your ripple state. Everything else, no matter how real, remains unread.

This work was not built in isolation. It was forged across conversations, classrooms, reflections, breakdowns, and breakthroughs.

It unites science with stillness, logic with delay, and action with non-interference.

If you are a physicist seeking closure, a philosopher seeking coherence, or a human simply seeking clarity—this principle offers a path. It won't demand belief. It will invite alignment.

This is not just about knowing the universe. It's about knowing when you're ready to read it.

Let the ripple begin.

— Sandeep Chavan

Independent Researcher, Educator, Founder of UED

How to Read This Book

This is not a book of equations—it is a book of realizations.

If you're expecting dense mathematics upfront, pause. This book doesn't begin with formulae—it ends with them. The math is not the premise; it is the **proof of perception**.

Here's how to approach it:

- **Start like a philosopher, not a physicist.** Let the metaphors, field concepts, and ripple analogies form naturally in your mind. This isn't intellectual decoration—it's how the truth reveals itself before it hardens into math.
- **Don't rush the uncertainty.** Much like the principle it redefines, understanding this book requires you to let insight align. You might not 'get it' instantly. That's the point.
- **Chapters build like ripples.** Each one begins with a small disturbance—a paradox, a misconception, or an ignored phenomenon—and gradually resolves into clarity. Let them resonate before moving on.
- **Pay attention to RAC and ATT.** These are not just terms—they are the book's gravitational centers. Everything else—from thought and vision to randomness and reality—swirls around them.
- **Read with your mind. Pause with your awareness.** Some sections are meant to be reread. Some equations may not click until the narrative loops back to them. That's ripple cognition at work.
- **Appendices are not optional.** They complete the narrative

arc by offering simulations, derivations, and real-world anchors to the philosophy. Read them like extended conversations.

This is a book for thinkers, teachers, researchers, and anyone who has ever looked at reality and whispered: *Something's missing...*

Don't try to master it. Let it ripple through you. *Because alignment is not a goal—it's a moment.*

Prologue

The Ripple That Refused to Align

A personal and scientific reflection on how uncertainty shaped understanding—and how ATT gave it meaning.

There was a time in my life when I thought I had lost everything—identity, purpose, and the faint sense of direction that once defined my pursuit of knowledge. That was 2011. A year where silence wasn't spiritual—it was suffocating. And uncertainty wasn't philosophical—it was existential. I was on the edge of disappearance, not from life, but from my own meaning.

But something refused to vanish.

Something small, intangible, yet persistent—like a ripple.

Not a scream, not a command, not even a clear thought. Just the subtle sensation that something in the universe knew I was still here, waiting to be heard, waiting to realign. At the time, I didn't have the language for it. Today, I do.

That something was a ripple. That refusal to vanish—was consequence. And that delay before clarity returned—was the *Alignment Threshold Time (ATT).*

Physics has long flirted with uncertainty. From Newton's precise laws to Einstein's elegant curvature of spacetime, science tried to pin the universe into a well-behaved box. But then came Heisenberg, like a storm that knocked over that box and said:

"You can't know everything. Not even in principle."

And we accepted it.

But we never really understood it.

We mathematized it. We diagrammed it. We quoted it. But we never stopped to ask:

"Why does uncertainty exist in the first place?"

Not the uncertainty of ignorance—but the kind that lives inside reality itself.

This book is not written in rebellion against Heisenberg. It is written as a completion of his intuition.

Heisenberg saw the curtain—but didn't name the actor behind it.

That actor, I believe, is consequence governed by alignment. And the missing law behind his principle is what I now call the *Ripple Alignment Criterion (RAC)*—a formula born not out of mathematics, but out of patience, field logic, and a personal battle with misalignment.

Through years of scientific exploration, failures, philosophical solitude, and rebirth of purpose, I realized:

The universe does not block truth. It delays it—until it is safe, stable, and ready to ripple back.

This delay is not uncertainty as randomness—it is uncertainty as recalibration. It is the ripple that refuses to align—not out of denial, but preparation.

That's what this book is about.

- Not about particles or probabilities.
- Not about collapsing wavefunctions or magical measurement.

It's about how consequence emerges from a field that listens before it responds. It's about how ATT became the most fundamental timer of truth—more than Planck's constant, more than speed of light, more than probability curves.

I write this now not just as a researcher. But as someone who has lived through the same alignment crisis that the ripple field undergoes when disturbed.

So, if you're reading this...

Know that uncertainty is not your enemy. It's the pause between your ripple and your realization. And ATT is not a number. It's the heartbeat of consequence.

Welcome to the field. It's still aligning. And this time—it's ready to ripple with your name.

Part I: The Problem with Uncertainty

Chapter 1: The Legacy of Heisenberg

What Heisenberg saw • Why it changed physics • Why it remained incomplete

"It wasn't that the universe stopped making sense. It simply stopped answering our questions with precision."

In 1927, Werner Heisenberg did something no physicist before him dared to do: He told the scientific world that certainty was a myth.

Not because our instruments were crude. Not because our minds were too small. But because reality itself refused to stand still long enough to be fully known.

With one equation—simple yet seismic—he drew a line across the legacy of classical physics. $\Delta x \cdot \Delta p \geq \hbar/2$ In that inequality, the illusion of control shattered.

From that moment, the world of particles, waves, and deterministic law collapsed into probability, fuzziness, and philosophical discomfort. The observer, once a quiet, passive measurer of truth, became a disturbing force—a ripple that altered the outcome by merely looking.

Heisenberg hadn't just discovered a quantum limitation. He had revealed the fragility of knowing.

And yet... for all its brilliance, something was missing.

What Heisenberg saw was real. What he felt was deep. But what he couldn't name... was the mechanism behind the uncertainty.

He left us with the truth of limitation, But not the reason for its arrival.

That's where our journey begins.

Because what if uncertainty wasn't a quantum glitch... But a ripple in transition? What if it wasn't random...But the consequence of the field still aligning?

And what if—just maybe—Heisenberg didn't end the conversation...But whispered the first question the universe was finally ready to answer?

The Crisis of Certainty

"The universe did not misbehave—it just refused to answer the wrong question."

For over two centuries, physics basked in the comfort of certainty. From Newton's gravitational elegance to Maxwell's electromagnetic equations, the universe appeared to be a grand machine—complex, yes, but ultimately predictable.

Every apple that fell. Every planet that spun. Every pendulum that swung.

They all whispered the same promise:

"If you know the laws, you know the future."

The world was understood in terms of perfect cause and effect. Given the initial conditions of a system, you could, in principle, calculate its entire trajectory into the future—or reconstruct its past. This wasn't just science. It was prophecy through mathematics.

At the heart of this predictive power stood classical mechanics. Time was absolute. Space was a fixed stage. Forces acted, masses responded. The human mind, it seemed, had finally unlocked the clockwork rhythm of the cosmos.

But Then... the Cracks Appeared.

As the 19th century faded and the 20th began, anomalies started to whisper from the shadows of experiments.

- **Mercury's orbit** deviated from Newtonian prediction.
- **The photoelectric effect** defied the wave theory of light.
- **Atomic spectra** displayed discrete jumps—not smooth

transitions.

- **Blackbody radiation** broke the rules of classical thermodynamics.
- **Light,** once so comfortably modeled as a wave, began behaving like a particle.

And just when scientists thought they could measure energy continuously, Planck's constant appeared—small, strange, and unignorable.

Light Itself Became a Paradox

It was neither purely a wave nor simply a particle.

- **Young's double-slit experiment** showed interference patterns—clear wave behavior.
- **Einstein's explanation** of the photoelectric effect proved photons deliver energy in quantized packets—clearly particle-like.

Physicists were caught in a philosophical cul-de-sac:

"How can the same entity behave as two contradictory things, depending on how we choose to look at it?"

This wasn't just a technical problem. It was a conceptual earthquake.

The Rise of Probability—and Discomfort

To make sense of this chaos, the early architects of quantum theory—Planck, Bohr, Einstein, Schrödinger, and Heisenberg—began to move away from determinism.

- They proposed quantization of energy levels.
- They embraced wavefunctions as probability maps.
- They introduced statistical mechanics as the language of systems too chaotic for prediction.

But this wasn't comfort—it was concession.

Science, once the art of knowing, was now the art of estimating. And the deeper one probed—especially at the atomic level—the more unpredictable reality became.

It was as if the universe had been playing along...until we looked too closely.

Enter the Observer

That's when the discomfort turned existential.

Suddenly, the mere act of observing seemed to alter the outcome. The observer was no longer a passive bystander—but a participant. And with that realization, the dream of full certainty collapsed.

"You can't just ask the universe questions anymore," said the equations. "You must consider whether it's ready to answer."

The Crisis, Defined

So, by the time Heisenberg arrived on the scene in the 1920s, physics was already in identity crisis.

The formulas still worked. But the confidence had evaporated.

It wasn't that the universe misbehaved—it simply refused to comply with outdated questions. It didn't break laws. It simply played by deeper ones we hadn't yet heard.

And thus began the true confrontation with uncertainty—not as human ignorance, but as reality's built-in refusal to rush alignment.

Heisenberg's Disruption

"You cannot know both where something is, and where it's going—not because you're unskilled, but because the universe won't let you."

The turning point came in 1927, when Werner Heisenberg, just 26 years old, dropped a conceptual bomb on physics:

You can't know everything—even in theory.

It wasn't just that our tools lacked precision, or that atoms were too small to observe clearly. Heisenberg went further. He claimed that nature itself forbade complete knowledge.

This was the moment where certainty died, and what we now call the Heisenberg Uncertainty Principle was born.

The Equation That Shook Physics

Heisenberg mathematically expressed his insight as:

$$\Delta x \cdot \Delta p \geq \hbar/2$$

Where:

- Δx is the uncertainty in position
- Δp is the uncertainty in momentum (mass × velocity)
- $\hbar$ is the reduced Planck's constant ($h/2\pi$)

At first glance, it looks like a basic inequality. But its meaning shattered centuries of deterministic thinking:

The more precisely you know where something is, the less precisely you can know where it's going—and vice versa.

This was not a limitation of instruments. This was a limit imposed by the universe itself.

Observation Is Not Passive

Before Heisenberg, it was assumed that observation could be *neutral*—that we could look at reality without disturbing it.

Heisenberg flipped that assumption.

He showed that at quantum scales, observation is always interaction. To "see" an electron's position, you must bounce something off it—like a photon. But doing so changes its momentum. And if you try to measure its momentum, you lose information about position.

It was as if the very act of watching created ripples that erased the clarity you were trying to capture.

Goodbye Determinism, Hello Probabilities

This realization wasn't just technical—it was deeply philosophical.

It meant:

- No matter how much data you collect...
- No matter how perfect your mathematics...

You can never predict the exact behavior of a particle.

The best you can do is predict the probability of where it might be found—or how it might behave.

This led to the rise of quantum mechanics as a statistical theory, not a deterministic one.

Reality was no longer a path to be traced. It was a field of possibilities to be interpreted.

The Copenhagen Interpretation: Probability as Truth

To deal with this uncomfortable truth, physicists turned to what became known as the Copenhagen interpretation, largely shaped by Niels Bohr and Max Born.

- **Max Born** proposed that the square of the wavefunction ($|\psi|^2$) gives the probability of finding a particle in a given state.
- **Bohr** added complementarity—the idea that particles can behave as waves or particles, but not both at once, depending on the setup.

In this framework:

- The observer defines the conditions of reality.
- Reality is not objective until it is measured.
- What exists before observation? No one could say.

And thus, certainty died—not just in practice, but in principle.

The Philosophical Recoil

Not everyone was happy.

Einstein famously rejected this approach, declaring:

"God does not play dice with the universe."

But Heisenberg—and the growing quantum consensus—stood firm.

Their message was clear:

"It's not about what we believe. It's about what the universe allows us to know."

Still, even as this principle became foundational, something remained unsettled. No one could explain why uncertainty existed. The math worked, but the mechanism was missing.

It was as if Heisenberg had mapped the edge of a ripple—but never saw the field beneath it.

Closing

What Heisenberg revealed was real. But the ripple was never fully named. The field that resisted our precision was never understood.

And without that understanding, the principle remained incomplete.

The universe wasn't playing dice. It was still aligning.

The Ripple He Didn't Name

"Heisenberg felt the resistance. But he never realized it was the ripple refusing to align."

Heisenberg's principle was powerful because it told us where we couldn't go. But it never told us why we were stopped.

The limitation was clear: the more precisely you measure one property, the less precisely you can measure its pair. But what exactly was resisting our measurement? What was *doing* the resisting?

This is where Heisenberg—like many great minds before him—touched something deeper than he could define.

He sensed the presence of a boundary...but could not locate its source.

Uncertainty as Consequence—but Without Cause

In physics, every principle has two parts:

- What it describes
- What causes it

Heisenberg gave us the description of uncertainty: a fundamental limit on simultaneous knowledge.

But he gave us no cause for that limit.

- Was it a feature of space itself?
- A glitch in the fabric of matter?
- A reflection of how energy moves through the vacuum?

These questions were never answered—only silenced under the umbrella of "it just is."

And so, uncertainty remained a floating truth—real, but incomplete.

The Missing Mechanism

What was absent in Heisenberg's theory was a mechanism—a physical process—that explained *how* and *why* uncertainty manifests.

There was no talk of:

- Time delay in the field's ability to stabilize a ripple.
- Field readiness or resistance to forced measurement.
- Ripple interference that could distort what was about to become a consequence.

Instead, the explanation was reduced to one abstract line: "Measurement disturbs the system."

But *why*?

What *exactly* is disturbed?

Heisenberg never answered. Nor did anyone else.

Because they weren't thinking in terms of the field.

Collapse or Misalignment?

Quantum mechanics told us that observation causes a wavefunction to collapse.

But collapse is a strange word. It implies something sudden, irreversible, and almost magical.

- Where did the rest of the wave go?
- Why did only one possibility survive?
- Why then? Why not later?

Again, no mechanism.

What was described as a collapse might actually be something else entirely:

A ripple that failed to align. A field consequence that couldn't stabilize.

But because there was no concept of ripple dynamics or structured vacuum equilibrium at the time, no one questioned collapse as an explanation.

They just accepted the result—and moved on.

Philosophical Discomfort

The more quantum theory was accepted, the more philosophers and physicists alike began to squirm.

- Did the universe really *not exist* until we measured it?
- Did electrons really lack position and momentum—until we forced them to declare one?
- Was reality itself fuzzy, or were our interpretations simply too crude?

Einstein, Schrödinger, and others raised these objections—but the math worked, and so the questions were sidelined.

Yet deep down, the discomfort lingered.

It was as if truth had been replaced by useful approximation. And science, in a moment of humility, gave up asking why, and settled for what works.

But as any true scientist knows, that's not the end. It's just the edge of the known.

A Field Unseen, A Ripple Unnamed

What Heisenberg *felt* in the heart of his principle was not just a limitation. It was resistance.

Not from logic, not from math—but from reality itself.

Heisenberg wasn't wrong. He was just early.

He sensed the ripple but never named it. He described the boundary but never defined its dynamics.

And so, uncertainty became a placeholder—a sign that something was happening beneath the surface of observation that our models had yet to uncover.

That *something* is what **Universal Energy Dynamics (UED)** now calls ripple misalignment. That limitation has a name: ATT—Alignment Threshold Time. And that incomplete principle has now found its next chapter.

Heisenberg gave us the ripple's shadow. UED now gives it light.

Why It Remained Incomplete

"Uncertainty wasn't the truth. It was a placeholder for something waiting to be defined. That something... is ATT."

Werner Heisenberg's insight was undeniably brilliant. He exposed a boundary where none had been acknowledged before. He forced physics to abandon its illusions of omniscience and perfection. He introduced a principle that humbled determinism and reshaped our understanding of knowledge itself.

But he never told us what lay on the other side of the boundary.

He drew the line. But he never crossed it.

And that is why his legacy—while foundational—remained incomplete.

A Boundary Without a Landscape

Heisenberg's equation revealed a limit to precision, but not the dynamics of that limit.

- Why does nature resist simultaneous knowledge of position and momentum?

- What internal process governs this trade-off?
- What *physically* happens when a ripple is observed?

Heisenberg gave us a mathematical curtain, but never looked behind it.

It was as if he'd discovered a pressure point in the universe, but didn't ask what was pressing back.

Planck's Constant: A Number Without a Narrative

The uncertainty principle relies on Planck's constant (ħ), a fixed value that sets the scale of quantum effects.

But what is ħ, fundamentally?

- It's not derived from first principles.
- It's not explained by any underlying mechanism.
- It simply *is*—a number inserted to make the equations match reality.

This reveals a deeper problem:

We have built entire theories atop constants we do not understand.

Planck's constant acts like a boundary marker—but no one ever asked why that boundary exists at all.

In UED, we replace that mysterious constant with something consequential: the Alignment Threshold Time (ATT)—a measurable delay that determines when ripples align into observable consequence.

The Absent Concept of ATT

Heisenberg's formulation lacked any notion of timing or field readiness.

There was:

- No concept of temporal alignment before consequence.
- No allowance for the field's memory or resistance.
- No parameter representing the delay between ripple initiation and response.

In essence, Heisenberg described the symptom of misalignment...but never suspected the existence of a minimum alignment time that governs all perception.

Had he conceived of ATT, he would've realized:

The universe doesn't reject simultaneous knowledge. It simply cannot support it within the same ripple window.

The Observer as Outsider

One of the central philosophical errors of quantum mechanics is its treatment of the observer as an external force.

Even in the Copenhagen interpretation:

- The observer "causes" the collapse.
- The observer "measures" the undefined.
- But the observer is never accounted for as a ripple in the same field.

Heisenberg's principle acknowledges that observation alters the system, but never explains how the observer is entangled with the ripple environment.

UED resolves this by declaring:

The observer is not an outsider. The observer is a ripple event within the same structured field. Therefore, their alignment (or misalignment) is the very cause of what is or isn't perceived.

Quantum Mechanics: Mathematically Effective, Philosophically Blind

Despite its predictive power, quantum mechanics has long struggled with its philosophical opacity.

It works. But it doesn't explain.

- Why does the wavefunction collapse? No answer.
- Why do probabilities become realities? No answer.
- Why does entanglement "know" faster than light? No answer.

These aren't side questions. They are the heart of science itself.

Heisenberg's principle sits at the core of this paradox: Accurate in prediction, empty in explanation.

That is not failure. That is a doorway.

Finishing the Thought He Started

Heisenberg opened the gate. But he never walked into the field.

What he called uncertainty, we now understand as hesitation—a ripple waiting to stabilize, a field still aligning, a universe not yet ready to respond.

And what he lacked—what the entire quantum framework still lacks—is ATT:

- The Alignment Threshold Time that governs when consequence can emerge.
- The Ripple Alignment Criterion (RAC) that determines when an observation becomes real.
- The clarity of consequence, not the blur of collapse.

Heisenberg saw the edge of the ripple. UED steps into its pattern and shows how it becomes reality.

Remember

"What Heisenberg discovered was not chaos—it was hesitation. And in that hesitation, the ripple waited for someone to finally name it."

Chapter 2: The Inheritance of Misunderstanding

Wavefunction collapse • Probability as a placeholder • The illusion of randomness

"Sometimes, the greatest mistake in science isn't what we don't know—It's what we assume too early, and then stop questioning."

After Heisenberg, the world didn't slow down to understand uncertainty. It rushed to institutionalize it. The boundary he discovered—the ripple he felt—was never investigated. It was packaged.

Wrapped in terms like *collapse*, *probability*, and *randomness*, the quantum revolution did something profound and dangerous:

It replaced depth with function.

And because the equations worked—because they predicted outcomes within statistical tolerance—the deeper questions were set aside.

The wavefunction became a mystical placeholder, treated more like a ghost than a ripple. Collapse was taught as a real physical event, despite no known mechanism. Probability, instead of a tool, was declared a law. And randomness... was accepted as a feature of the universe, rather than a consequence of not waiting for alignment.

This wasn't deception. It was urgency turned into dogma. It was physics choosing prediction over understanding, and calling the fog *truth*.

But here's the uncomfortable truth:

The wavefunction never collapsed. The universe was never random. And probability was never real—it was our impatience made scientific.

What we inherited from early quantum theory was not a complete picture. It was a brilliant sketch—abandoned mid-stroke.

This chapter reopens that sketch, wipes the fog from its outlines, and asks:

- What really happens during measurement?
- Why does randomness appear at all?
- And how did misunderstanding become canon?

It's time to reframe collapse as failed alignment, probability as ripple blindness, and randomness as consequence without readiness.

Let's begin.

The Collapse That Never Was

"The wavefunction didn't collapse—it simply failed to align into consequence."

In the early 20th century, the wavefunction (ψ) emerged as one of the most iconic symbols of modern physics. Encoded within it was everything a quantum system could potentially be—every state, every path, every energy level.

But unlike classical variables, the wavefunction wasn't about what *is*. It was about what *might be*.

It didn't give certainty. It gave a cloud—a diffuse, probabilistic haze over reality. A soft glow of possibility... awaiting collapse.

From Possibility to a Single Outcome

But here was the problem: In actual experiments, we don't see a cloud. We see a click on a detector. A dot on a screen. A single, definite outcome.

So, physicists needed a bridge between:

- The many possibilities in the wavefunction before observation

- The one outcome seen after observation

Enter the concept of wavefunction collapse.

Collapse: A Concept Born From Discomfort

According to standard quantum mechanics:

- Before measurement, the particle exists in a superposition of all possible states.
- During measurement, the wavefunction instantaneously collapses into one outcome.
- The rest of the possibilities? They vanish.

But what does it mean to collapse a wavefunction?

- Does the universe "decide" which version to keep?
- Where do the other possibilities go?
- Is collapse a physical event? Or just a mental update?

The theory had no mechanism, no time delay, no causal structure.

And yet, collapse was adopted as law, despite being purely interpretative.

It wasn't the wavefunction that collapsed. It was our understanding that folded under pressure.

Schrödinger's Cat: Collapse Made Absurd

To highlight the absurdity of the collapse concept, Erwin Schrödinger proposed a now-famous thought experiment:

- A cat is placed in a box with a quantum mechanism that has a 50% chance of releasing poison.
- Until observed, the system is said to be in a superposition of "dead cat" and "alive cat."
- Once someone opens the box, the wavefunction collapses—and the cat is either alive or dead.

This isn't science. It's satire.

Schrödinger was showing that if we take collapse literally, we must accept the cat is both alive and dead until a conscious being looks inside.

Collapse, in this light, becomes an intellectual crutch—a forced fix to make the math align with the consequence.

And that's the key:

Collapse was never an event. It was a narrative patch.

Measurement: The Black Box of Modern Physics

In all quantum formulations, measurement is treated as a black box:

- *Before*: The system exists as overlapping probabilities.
- *After*: One definite outcome is recorded.
- *How* this happens? Left unanswered.

No account of:

- What physical process "selects" the outcome
- How the rest of the wavefunction disappears
- Whether the field responds gradually, or not at all

It was easier to say:

"The wavefunction collapsed." Than to admit: "We don't know what happens in between."

In UED, we challenge that silence.

We say:

There is no collapse. There is only ripple interference and misalignment. If the observer is misaligned with the ripple field's state—consequence becomes fuzzy or delayed.

But when the ripple aligns—with sufficient ATT, amplitude, and phase coherence—a consequence locks in.

Not randomly. Not magically. But consequentially.

Insight Reframed

Collapse is not a fundamental phenomenon. It's an interpretation born from an incomplete field model. It is what you invent when you lack the language of ripple timing and field readiness.

In truth:

The wavefunction doesn't collapse. It fails—or succeeds—to align into consequence. And that alignment is governed not by chance, but by ATT—the universe's way of saying: *"Not yet. I'm still preparing your answer."*

Probability—The Language of Ignorance

"Probability is not the truth—it's what we say when we don't know the alignment conditions."

In 1926, Max Born proposed a revolutionary interpretation of Schrödinger's wavefunction (ψ):

The square of its magnitude—$|\psi|^2$—does not describe physical reality itself, but gives the probability of finding a particle at a certain position.

This wasn't just a mathematical suggestion. It became the default belief system of quantum mechanics.

No longer was the wavefunction a representation of what is. It became a statistical map of what might be.

From Possibility to Probability

Before Born's interpretation, there was still a faint hope that quantum theory could yield a deeper, deterministic picture. That the wavefunction might represent real, physical vibrations of some hidden medium.

But with Born's idea, that hope faded.

The Copenhagen interpretation now said:

- Particles don't have definite properties until measured.
- Until then, we speak in terms of probabilities.

And so began a new chapter in physics—one that replaced missing causes with probability distributions.

But Probability of What?

This is the uncomfortable question that was never fully addressed:

- What exactly is this probability a measure *of*?
- Is it a property of the particle?
- Of the observer?
- Of the field conditions?

Or is it simply a measure of how blind we are to what's really happening?

Instead of asking these deeper questions, physics simply rebranded ignorance as law.

Probability as a Substitute for Missing Causality

In the absence of a mechanism for how outcomes emerge from the wavefunction, probability became the patch.

- No understanding of how interference stabilizes? → Use probability.
- No idea why a measurement yields a specific result? → Use probability.
- Can't explain collapse? → Assign a distribution and move on.

This is not science. It's statistical surrender.

Probability in quantum mechanics doesn't reflect the universe's randomness. It reflects our refusal to model alignment.

Ontological Randomness vs Epistemic Uncertainty

Let's draw an important distinction here.

There are two kinds of uncertainty in philosophical and scientific discourse:

Type	Meaning	UED View
Ontological randomness	Reality is inherently random. Events happen without cause.	Rejected. Ripple fields are consequential, not chaotic.
Epistemic uncertainty	The system behaves consistently, but we lack the information to predict it.	Accepted. Misalignment of observer with ripple state = uncertainty.

Quantum physics, by adopting Born's probability as fundamental, slid toward ontological randomness—as if nature flips coins behind our backs.

But UED reclaims the truth:

The universe is not flipping coins. It's rippling toward equilibrium—and we often try to measure it before it's done.

The Danger of Canonizing Ignorance

When probability is used as a temporary bridge, it's helpful.

When it becomes a law, it's dangerous.

It shifts science from curiosity to compliance. It tells generations of physicists:

- "Don't ask why."
- "Don't expect cause."
- "Just accept the numbers and run the simulations."

This mindset is what UED directly challenges.

Because once we account for:

- ATT (Alignment Threshold Time)
- Phase interference ($\Delta\Phi$)
- Observer amplitude (A_o)

...we realize that what looked probabilistic was simply unresolved ripple dynamics.

Probability wasn't telling the truth. It was covering for our blindness.

Insight Reaffirmed

Probability is not the voice of nature. It's the whisper of our own impatience.

We use it when we haven't yet learned how the field is behaving, or when we refuse to ask what the ripple is really doing.

Born gave us a mathematical tool. But generations turned it into a philosophical wall.

And now, it's time to tear that wall down—by understanding what determines when and how ripple alignment leads to consequence.

The Myth of Randomness

"Randomness is how reality behaves when you try to measure it while it's still thinking."

Few words in modern physics carry more philosophical weight—and more quiet confusion—than "randomness."

Quantum theory, as it stands today, insists that randomness is real:

- That the precise outcome of any quantum measurement is inherently unpredictable.
- That nature makes no choice until the observer looks, and when it does, it chooses arbitrarily.
- That this randomness is not a temporary gap in knowledge—it is a feature of reality itself.

But Universal Energy Dynamics (UED) takes a hard position here:

There is no such thing as true randomness. What appears random is merely a reflection of misalignment, interference, or incomplete ripple recognition.

The Legacy Claim: Randomness as a Built-in Feature

Textbooks teach that a quantum particle's position, spin, or momentum is undefined until measured—and when measured, its final state is selected randomly from the possible outcomes governed by $|\psi|^2$.

Even Einstein, who fundamentally resisted this interpretation, eventually stood alone. His discomfort—*"God does not play dice"*—was dismissed as philosophical conservatism.

Over time, randomness was canonized:

- It became a metaphysical excuse for uncertainty.
- A scientific mask for unexplained ripple transitions.
- A way to shut down inquiry, not deepen it.

And it worked—mathematically. But it never worked philosophically, causally, or consequentially.

UED Response: No Randomness, Only Ripple Misalignment

UED introduces a completely different lens:

- Every quantum system exists within a structured vacuum ripple field.
- Every "event" or consequence is the result of ripples aligning in a measurable, resolvable way.
- What we call randomness is simply an observer attempting to extract consequence when:
    - The field has not completed realignment,
    - The observer's ripple amplitude or timing is off, or
    - The phase interference within the local field distorts the consequence.

There is no dice roll. There is only ripples refusing to converge in that moment.

Randomness is a field's polite way of saying: "I'm not ready yet."

Randomness as a Symptom, Not a Principle

Let's reframe randomness:

- It is not an explanation.
- It is a signal that something deeper is not being measured.

In legacy science, randomness is declared a primitive truth. In UED, it is treated as diagnostic feedback:

Symptom	Underlying Cause
Apparent randomness	Misalignment of ripple timing or phase (ΔΦ)
Unstable measurement	ATT not satisfied
Unpredictable consequence	Observer's ripple amplitude too weak or distorted

Just like noise in a radio doesn't mean music is random—it means you're not tuned to the right frequency yet.

ATT and RAC: Tools to Deconstruct Randomness

UED gives us two concrete tools to *remove the illusion of randomness*:

ATT (Alignment Threshold Time)

- Defines the minimum duration the field needs to allow consequence to manifest.
- If measurement occurs before ATT is met, the result appears random—because the field was still realigning.

RAC (Ripple Alignment Criterion)

- Combines observer amplitude, available duration, and phase shift into one testable inequality.
- If RAC is not satisfied, alignment fails, and consequence becomes unstable.

$$(A_o/A_r) \cdot \Delta\tau_{obs} \cdot \cos(\Delta\Phi) < \mathrm{ATT}_{min} \Rightarrow \text{Perceived randomness}$$

This isn't mystical. It's causal. It's empirical. And most importantly—it's explainable.

The Real Danger of "Randomness"

By accepting randomness as a feature, we do three things:

1. We stop looking for cause. → Inquiry halts at probability.
2. We disconnect human perception from cosmic structure. → As if the observer is uninvited, unentangled, unimportant.
3. We mistake noise for nature. → When in fact, nature is always coherent—if you wait long enough.

Insight Reaffirmed

Randomness is not what the universe gives us. It's what we perceive when we interrupt before consequence is born.

It's the symptom of a measurement made too early, Of an observer misaligned, Of a field still weaving the ripple into form.

UED doesn't deny randomness. It disassembles it.

And in the pieces, we find a deeper rhythm—one where consequence is not uncertain... Just untimed.

Institutionalized Misunderstanding

"Quantum mechanics didn't lie to us—it just left the conversation too early."

The early quantum pioneers—Planck, Heisenberg, Bohr, Born, Schrödinger—stood at the edge of a monumental discovery. They glimpsed a new reality pulsing with probabilistic behaviors, measurement effects, and superposed states. But rather than walking further into the unknown, science made a subtle but fateful decision:

It chose mathematical success over philosophical clarity. It chose calculation over causation. It chose to close the case—just as the most important questions were about to begin.

From Insight to Dogma

Initially, wavefunction collapse, superposition, and probability were working assumptions—scaffolding to hold together a strange but functional theory. They were never meant to be final truths. Even Bohr and Born admitted that their ideas were interpretive tools, not deep ontological claims.

But the more these ideas produced correct numerical predictions, the more they were mistaken for truth itself.

- Collapse wasn't just a placeholder—it became a law.
- Probability wasn't just a tool—it became reality's foundation.
- Randomness wasn't just a pattern—it became a principle.

The scaffolding was never removed. It was painted over, institutionalized, and enshrined in textbooks.

The "Shut Up and Calculate" Mindset

As quantum mechanics matured and fueled technological breakthroughs—lasers, transistors, semiconductors, quantum computing—something else began to decay: the hunger for explanation.

Instead of pursuing the *why*, physicists were told to focus on the *what works*. The now-infamous mantra echoed across labs and classrooms:

"Shut up and calculate."

This attitude didn't emerge out of laziness. It emerged from exhaustion. The philosophical problems of quantum theory were too uncomfortable, too paradoxical, and—critically—they didn't stop the math from working.

So the field retreated from inquiry and focused on application. And in doing so, it quietly normalized incompleteness.

Teaching the Tools as Truth

By mid-20th century, the Copenhagen interpretation wasn't just the dominant school of thought—it was taught as reality itself.

Students learned:

- Collapse is what actually happens.
- Probability is how nature functions.
- The observer triggers consequence.

What began as conceptual scaffolding became institutionalized misunderstanding.

The language of early caution—*"we think," "it appears," "as if"*—was dropped. What remained were bold statements disguised as settled facts.

And over time, generations were raised inside this frozen narrative.

No one questioned why the wavefunction collapses. They were simply taught that it does.

No one asked what causes randomness. They were taught it's baked into nature.

Consequences of a Closed Model

This premature closure had ripple effects far beyond physics:

1. Science lost its humility.

A discipline once rooted in open questioning became content with partial answers.

2. Philosophy was shut out.

The metaphysical implications were abandoned, dismissed as "unscientific"—even though the heart of the uncertainty principle is philosophical.

3. Public understanding fractured.

People were told the universe is random, unstable, and unknowable. This seeded cultural cynicism, spiritual confusion, and intellectual paralysis.

4. Future generations inherited a wall, not a window.

Students weren't taught to question foundational assumptions. They were taught to manipulate equations that had no intuitive grounding.

UED's Re-entry into the Conversation

UED doesn't come to destroy quantum theory. It comes to continue the conversation quantum theory walked away from.

It says:

- Collapse is not necessary.
- Probability is not fundamental.
- Randomness is not real.

These were symptoms of a field that was still aligning, a ripple that hadn't yet stabilized, a consequence that hadn't ripened.

And what we inherited as "law" was only a description of what we didn't wait to understand.

Insight Reaffirmed:

"Quantum mechanics didn't lie to us—it just left the conversation too early." "It gave us the boundaries. UED gives us the mechanism." "It showed us the blur. UED shows us the ripple beneath it."

Remember

The first generation of quantum physics gave us tools that worked. But they were never meant to be the final truth. They were signals. Pauses. Moments where the universe said:

"There's more. But you have to stay longer to hear it."

Now, with ATT and the Ripple Alignment Criterion, we return to those unfinished thoughts—and carry them forward, from probability to consequence, from collapse to clarity.

The Inheritance of Misunderstanding

This chapter began by revisiting the deeply embedded concepts of wavefunction collapse, probability, and randomness—not to deny their historical usefulness, but to expose their philosophical shallowness. What emerged from our analysis is clear: the quantum revolution solved equations, but left questions unanswered. And in that gap, misunderstandings calcified into doctrine.

The Wavefunction Never Collapses

Contrary to what textbooks preach, the wavefunction doesn't collapse as a physical act. It doesn't spontaneously reduce, vanish, or snap into reality.

What actually happens—when observed from the lens of UED—is one of two things:

- The ripple field aligns into measurable consequence,
- Or it fails to, due to insufficient conditions (phase interference, timing, amplitude mismatch).

The phenomenon we call "collapse" is just a label for this real-time consequence resolution, governed by a minimum alignment delay: the ATT (Alignment Threshold Time).

Collapse was never a law—it was a sign that we lacked a better vocabulary for consequence in transition.

Probability Is Not Law—It's Linguistic Survival

Probability wasn't discovered—it was adopted. It was how we made peace with incomplete awareness of the ripple environment.

Born's interpretation of $|\psi|^2$ gave us a statistical scaffold. But over time, this mathematical convenience became confused for ontological truth—as if the universe itself operates through dice rolls.

In UED, probability becomes a diagnostic tool, not a fundamental principle.

When we don't understand ripple interference or ATT conditions, we speak in probabilities. But once we do, outcomes become explainable, not just predictable.

Randomness Is the Shadow of Misalignment

What physics called "randomness" was simply the visible symptom of invisible dynamics.

- A measurement made before the field has stabilized.
- An observation attempted without coherent phase alignment.
- An observer unaware of their own ripple's influence.

These are not mysteries—they are misreadings of readiness.

The field isn't random. It's just not yet responding in a way we can detect.

In this sense, randomness is our projection—the flicker on the screen when the signal hasn't locked.

Partial Truths, Taught as Whole Explanations

The architects of early quantum theory never meant for these concepts to be taken as final. They were reaching, not concluding.

But as their equations began to succeed technologically, curiosity took a back seat to efficiency. Collapse, probability, and randomness were left on the throne, not because they were flawless, but because they were functional.

What we inherited, then, wasn't a lie. It was a half-lit map, mistaken for the territory.

Closing Insight:

"They taught us to calculate uncertainty. Now, we're ready to understand it."

We're no longer content with math that works. We seek meaning that explains. With ATT and the Ripple Alignment Criterion, we don't abandon quantum mechanics—we complete it.

We don't replace precision with poetry—we restore consequence to the center of reality. And most of all, we reclaim understanding from the edge of silence.

Chapter 3: The Observer Crisis

How observation became a villain • The passive vs active observer • When did science stop listening?

"Once upon a time, science trusted the one who watched. Then quantum mechanics arrived—and the watcher became the problem."

In classical physics, the observer was silent. You could measure the trajectory of a planet or the motion of a pendulum without fear that your presence would corrupt the system. You were simply *there*—a passive witness to a predictable universe.

But quantum theory shattered that trust.

Suddenly, the act of observation wasn't neutral—it was dangerous.

- It could collapse a wavefunction.
- It could destroy superposition.
- It could create a reality that hadn't yet existed.

From passive to powerful. From quiet to corruptive. From witness to villain.

This shift wasn't just a mathematical wrinkle. It was a philosophical earthquake. Because for the first time in history, science told us:

"Reality doesn't exist the way you think—until you interfere with it."

And so began the observer crisis—A crisis not just of physics, but of perspective.

- What is the observer, really?

- Do we shape the world simply by looking at it?
- Or are we just measuring echoes of our own disturbance?

Worse, in all this uncertainty, one question faded into silence:

"When did science stop listening to the observer—not as a problem, but as part of the field?"

This chapter explores how observation was misunderstood, misframed, and ultimately blamed for mysteries that ripple dynamics could have explained all along.

We will dismantle the false duality of passive vs active, and reveal the truth: the observer is not an intruder—they're a ripple with intent, bound by the same rules of alignment and consequence as anything else.

The problem was never observation. The problem was our refusal to understand what it means to observe.

Let's begin.

From Witness to Threat – The Villainization of Observation

"The observer didn't disturb nature. It was nature trying to respond to the observer that disturbed our models."

For most of scientific history, the observer was a silent spectator—someone who watched, measured, recorded, and remained irrelevant to the system being observed.

In classical physics, this was a cornerstone assumption:

- You could track a planet without changing its path.
- You could record a falling object without slowing it down.
- You could measure a phenomenon without being part of it.

The observer was external. The system was autonomous. Reality unfolded, and the observer simply took notes.

This belief—so fundamental to Newtonian and relativistic frameworks—gave birth to objective science: the idea that nature has a state that exists, regardless of who's looking or whether anyone is looking at all.

But then came quantum mechanics.

And suddenly, the observer wasn't a witness anymore. They were a variable.

Observation Became a Disturbance

In the quantum world, observation did not just reveal reality—it interfered with it.

- To detect an electron's position, you had to bounce a photon off it—changing its momentum.
- To measure spin, you had to pass it through a field that altered its alignment.
- To "observe" the wavefunction, you had to collapse it.

What had once been a neutral act now became a contaminating one.

And so, the observer—formerly a scientist's greatest ally—was rebranded as a threat.

"The moment you look," said quantum theory, "you change what you see."

The Measurement Paradox

This shift gave rise to one of the most bizarre paradoxes in modern science:

You cannot describe what is—because the act of describing changes what is.

The act of measurement, instead of being a tool for accessing truth, became a barrier to it. And the deeper you tried to measure, the less certain the truth became.

This broke science's confidence.

- If observation *causes* outcomes, can we trust any result?
- If reality isn't stable until we measure it, what is it before we look?
- Is the universe waiting on us to decide what happens?

Suddenly, physics became entangled in metaphysics—and not in a good way.

The Dual Tragedy

This led to two simultaneous tragedies in modern science:

1. Observation became feared.

Instead of a bridge to understanding, it became a disturbance to be minimized, avoided, or "mathematically isolated."

2. The observer became misunderstood.

Reduced to a "problem," the observer was never properly modeled—not as a physical entity, not as a field interaction, and certainly not as a ripple participant in the same system they were trying to access.

In other words, science turned against itself. The very act that makes science possible—observation—was now seen as the reason science cannot fully work.

This was never a failure of nature. It was a failure of framing.

The UED Correction

UED does not reject the importance of the observer. It redefines it.

The observer is not a separate being, outside the field. The observer is a ripple—one that emerges within the same structured vacuum as the phenomena it tries to perceive.

When you observe something, you are not disturbing the system. You are initiating a ripple alignment sequence.

If ATT (Alignment Threshold Time) is not met, and if phase interference ($\Delta\Phi$) is high, your observation yields distortion. But if conditions align, consequence stabilizes—and observation becomes resonance, not interference.

It was never the observer that disturbed nature. It was nature trying to respond to the observer that made our outdated models collapse.

Insight Reaffirmed:

- Classical science made the observer invisible.
- Quantum science made the observer intolerable.
- UED makes the observer integral.

Not because they cause the event... But because they complete the ripple Criterion necessary for consequence to emerge.

The Passive vs Active Observer

"The observer doesn't cause consequence. The observer becomes part of the condition that allows it."

Throughout the evolution of physics, the role of the observer has undergone a profound identity crisis. From the invisible witness of Newtonian mechanics to the indispensable actor of quantum mechanics, the observer has shifted from a mere chronicler of events to a participant in the very formation of reality.

But the confusion didn't just lie in these evolving roles—it lay in the frameworks that couldn't explain them.

Let's break it down.

The Passive Observer: Classical Fantasy

In classical physics, the observer was treated as a neutral frame of reference. Observation was assumed to be non-intrusive—like watching rain through a window without changing its path.

- A ball thrown followed a predictable arc, regardless of who watched it.
- A planet orbited the sun, with no regard for telescopic eyes on Earth.
- Light, sound, temperature—all could be measured without altering the system.

This created the myth of the passive observer:

An outsider, unaffected and uninvolved.

This worked for centuries. But it was never truly accurate. It simply didn't matter—until systems got small, fast, and sensitive enough to reveal the flaw.

The Active Observer: Quantum Panic

Quantum physics smashed the illusion of the passive observer. Suddenly:

- You could not observe a system without being part of it.
- You could not measure a property without influencing it.
- You could not even define an outcome without the measurement itself collapsing all other possibilities.

This gave rise to the active observer—not by virtue of choice, but by necessity.

In this view, the observer is entangled, embedded, inescapable.

But while this was a more accurate description of the phenomena, it came with no clear mechanism. Why did observation "collapse" the wavefunction? Why did measurement seem to dictate reality?

No one knew.

So, physicists built mathematical workarounds. They spoke in terms of probability, collapse, expectation values, and statistical ensembles.

But the observer remained an uncomfortable ghost in the machine—always present, never explained.

UED's Framing: The Observer as Ripple

UED dissolves the passive–active duality entirely. In Universal Energy Dynamics:

The observer is not outside the field. The observer is a ripple within it.

Observation is not an intrusion. It is a ripple alignment attempt.

Every conscious act—seeing, measuring, questioning—is an energetic ripple trying to align with the existing field.

That alignment can succeed or fail depending on two key conditions:

ATT (Alignment Threshold Time)

- The delay required for a ripple to properly synchronize with the field's dynamics.
- If ATT is respected, observation is possible. If rushed, it's distorted.

ΔΦ (Phase Coherence)

- The degree to which the observer's ripple is in phase harmony with the field.
- Misalignment leads to noise, collapse illusions, or randomness.

Hence, the observer is not passive (uninvolved) Nor is the observer active (disruptive)

The observer is conditional. A ripple seeking coherence with reality's current ripple field.

And when coherence is achieved? Observation becomes consequence. Reality unfolds—not because it was forced, but because the conditions were ripe for it to appear.

The Ripple Insight Reframed

"The observer doesn't cause the event. They become part of the ripple choreography that allows it to happen."

This means:

- Observation is not a button that triggers collapse.
- It is an invitation, accepted only if ripple conditions permit.
- The field is not resisting the observer.

- The field is waiting—for ripple harmony to be established.

A New Role for the Observer

Rather than villainizing the observer or worshiping their "power," UED reframes the observer as an alignment instrument.

Their clarity, readiness, and patience determine whether they witness distortion or truth.

This transforms the observer from a philosophical dilemma...into an energetic participant in the symphony of consequence.

The Collapse of Listening in Science

"Science didn't just stop listening to the field. It stopped listening to itself."

There was a time when science was not just inquiry—but introspection. It asked *why*, not just *what*. It revered the mystery as much as the model.

But somewhere between the equations of Schrödinger and the successes of silicon, something changed.

When Did Science Stop Asking "Why"?

In the early 20th century, science was alive with wonder. Einstein asked: *What is time?* Bohr asked: *What can we know?* Heisenberg asked: *What does it mean to observe?*

There was room for doubt, for philosophy, for interpretation. But when quantum mechanics began producing results, a different voice took over.

That voice didn't want to explore—it wanted to predict. It didn't want truth—it wanted utility.

Thus began the era of:

"Shut up and calculate."

Coined not as a joke, but as a creed.

This mindset sidelined metaphysical discomfort in favor of mathematical consistency. It rejected the observer not because it had understood the role—but because it couldn't model it neatly.

The Rise of Blind Calculation

In this era, questions about the observer's role, intelligence, or alignment were deemed irrelevant.

What mattered was:

- Can we predict the electron's probable location?
- Can we compute the energy levels?
- Can we make a better transistor?

And to be fair—science thrived technologically.

But philosophically? It went mute.

The observer was reduced to a coordinate system, a measuring stick, a clock. A non-entity in a universe supposedly built to be observed.

What Was Ignored?

In pushing the observer to the sidelines, science ignored something far more consequential:

- The quality of observation
- The state of the mind doing the measuring
- The clarity, awareness, and intent of the one interacting with the field

It never asked:

- Is the observer aligned?
- Is the observer's ripple stable?
- Can perception vary based on internal coherence?

The answer is yes. But science was no longer listening.

What This Meant for Philosophy and Spirit

In silencing the observer, we silenced:

- Self-inquiry
- Spiritual questions about interconnectedness
- Cognitive science on awareness and mental readiness

This wasn't just a scientific error—it was a civilizational loss.

By removing the observer's depth, science declared that:

Your consciousness is irrelevant. Your presence is noise. Your curiosity is a distraction. And in doing so, it amputated its own soul.

UED's Restoration of Listening

UED reopens the conversation. It reminds us that:

- The field speaks—but only in ripples.
- Consequence unfolds—but only when alignment permits.
- Observation matters—but only when the observer is in tune.

In UED, the observer is not just a participant—they are a condition.

And if their ripple is unclear, anxious, or misaligned, the field will respond with distortion, delay, or ambiguity.

Not because reality is hiding— but because it is waiting to be met by someone who can hear it.

Closing Reflection:

What science once called randomness may be the field whispering in a frequency too subtle to catch.

What it called collapse may be ripple fatigue from impatient measurement.

What it ignored—was itself.

"When the field went silent, it wasn't because it had nothing to say. It was because science stopped listening to the way it speaks."

UED's Observer – A Ripple-Aware Consciousness

"To observe is not to extract—but to resonate."

The observer is not an outsider. Not a detached machine. Not a passive clock counting ticks.

In Universal Energy Dynamics (UED), the observer is something far more profound:

A ripple-aware interface—a conscious field pattern attempting alignment with the larger ripple field of reality.

This changes everything.

Beyond the Measuring Device

In mainstream science, the observer is often reduced to instrumentation:

- A camera records.
- A sensor detects.
- A lab technician notes the reading.

But these tools are not observers—they are extensions of the observer's will. They do not ask. They do not seek. They do not resonate. They merely record after the ripple has spoken—or misfired.

In UED, observation is not detection. It is interaction. It is co-resonance between observer and field.

Observation = Participation, Not Interruption

Where quantum mechanics feared the observer—worried that looking destroyed the system—UED corrects the misunderstanding:

Observation doesn't destroy the ripple. Impatience does.

Observation is not an intrusion. It is a request for resonance.

The field does not collapse—it responds conditionally.

The Role of ATT – Alignment Threshold Time

ATT is the minimum time required for the observer's ripple to stabilize and align with the ongoing ripple environment.

If ATT is not respected, the following occurs:

- The observer's ripple interferes with the system prematurely.
- Distorted signals emerge—perceived as randomness.

- The field appears chaotic, simply because it wasn't ready to respond.

Thus, ATT defines *when* observation is energetically permitted.

It's not about speed. It's about synchrony.

The Role of RAC – Ripple Alignment Criterion

RAC determines how well the observer's ripple matches the phase, amplitude, and orientation of the field ripples it tries to observe.

RAC is not about closeness. It's about coherence.

- **Perfect RAC** → Stable, repeatable consequence.
- **Weak RAC** → Fuzzy outcomes, inconsistent results.
- **No RAC** → No observable event, or full misinterpretation.

Together, ATT and RAC define the possibility and quality of observation.

When Alignment Succeeds...

When an observer respects ATT and meets RAC, the result is profound:

- **Consequence becomes clear**: The field responds precisely.
- **Measurements are consistent**: Not because the world is predictable, but because the observer is aligned.
- **Events are meaningful**: Perception matches reality, not illusion.

This is not magic. It is ripple choreography—the field offering itself to the ripple that dances in harmony.

When Alignment Fails...

On the other hand:

- **ATT is rushed** → The observer's ripple slams into an unready field.

- **RAC is poor** → Misaligned perception, fuzzy or contradictory outcomes.
- **Observation is attempted in noise** → The system echoes confusion.

What we call quantum randomness, wavefunction collapse, or statistical variance—these are all symptoms of failed alignment.

The field didn't break. The observer wasn't yet part of the rhythm.

The Conscious Observer

In UED, the observer is not just physical—it is aware.

Consciousness is not an emergent property. It is a ripple stabilization function.

- It helps refine RAC through intent and attention.
- It holds ATT patiently, allowing time for resonance.
- It discerns when consequence is authentic—and when it's noise.

Observation, then, becomes a discipline, not a button.

It becomes a relational act, not a transaction.

It becomes a spiritual skill, not just a scientific step.

Final Realization

Observation is not an act of extracting reality from nature. It is an act of tuning yourself to the ripple conditions that allow truth to emerge.

"To observe is not to extract—but to resonate."

The observer is not a variable in the equation. They are the field in form, the ripple in readiness, the alignment in motion.

And when the observer truly aligns?

The universe does not just show itself—it welcomes you home.

The Observer Crisis

The fall from grace of the observer in modern physics wasn't a result of failure—it was the consequence of misunderstanding ripple dynamics.

We began with classical physics, where the observer was irrelevant, external, and non-interfering. The world existed *out there*, unfolding in predictable harmony. The act of looking was innocent.

But with the quantum revolution came a new anxiety: observation alters the observed. Heisenberg's uncertainty. Schrödinger's cat. The wavefunction collapse. Suddenly, to observe was to ruin.

Physics panicked—and in that panic, it committed an ironic act:

It overdefined the observer while understanding them less.

We weren't ready to ask what an observer truly is. We just blamed them for what we couldn't yet explain.

What Went Wrong?

- The observer was never meant to be passive—but quantum mechanics made them too active, too soon.
- We treated measurement failure as a collapse of reality, rather than a failure of alignment.
- We mistook field unreadiness for nature's chaos.
- We tried to calculate without listening—and ended up calling the field's silence a randomness.

In truth, the field wasn't collapsing. It was simply not yet aligned with the observer.

What UED Restores

UED reframes this crisis—not as a quantum anomaly, but as a ripple miscommunication.

- The observer is a ripple: not a foreign agent, but a localized expression of the same field.
- Observation is resonance, not extraction.
- ATT (Alignment Threshold Time) tells us when the field is

ready.

- RAC (Ripple Alignment Criterion) tells us how clearly we can engage with it.
- Observation succeeds not through precision, but through readiness, patience, and alignment.

The act of observing is no longer a risk—it is a responsibility. It is no longer about taking—it is about *tuning in.*

Final Insight

We didn't have an observer problem. We had a listening problem.

"The observer isn't the one who disturbs the universe. They're the one who must wait for its response to complete."

In the silence before consequence, in the space between perception and participation, the ripple awaits clarity.

And once again, physics doesn't need to collapse its questions—It just needs to ask them differently.

Interlude: Between Collapse and Clarity

We have walked through the storm of misunderstanding.

From Heisenberg's cautious genius, to a century of partial truths, to the silent exile of the observer—uncertainty has cast a long shadow over modern physics.

But now, something is shifting. We've seen that what we called "collapse" was never a fall—but a hesitation. That "randomness" is not a property of nature, but a symptom of impatience. That the observer was never the enemy of truth—only the one left out of the conversation.

Physics didn't lie. It simply didn't wait long enough to hear the full answer. This is the space we now occupy. The space between old interpretations and a new understanding.

The collapse narrative has run its course. It gave us impressive predictions, but left us with deeper questions. It told us how—but not why. It offered equations—but not coherence.

Now, with UED, we don't abandon the past. We realign it.

We introduce the idea that ripple dynamics—governed by ATT (Alignment Threshold Time) and RAC (Ripple Alignment Criterion)—offer not a contradiction to uncertainty, but a completion.

Not a rejection of quantum mechanics, but the beginning of its reconciliation with consequence, clarity, and consciousness.

"We're not escaping uncertainty. We're revealing its structure."

As we step into Part II, we are no longer merely critiquing. We are constructing. The hesitation Heisenberg named will now find its rhythm. The observer's silence will now become a signal. And the ripple—no longer a metaphor, but a measurable dance of alignment—will finally speak.

Part II: Uncertainty Reborn in UED

Chapter 4: Ripple Fields

A New Medium of Reality

We searched the universe for a hidden force. We split atoms, bent spacetime, invoked particles from nothing, and tried to catch certainty in probability nets. But the more we looked, the less we understood. Everything—matter, motion, even light—refused to behave when we tried to pin it down.

Maybe it was never a behavior problem. Maybe it was a medium misunderstanding.

Modern physics treated the vacuum as empty. Quantum physics treated it as buzzing noise. But neither asked the most important question:

What if the vacuum is not empty or noisy—but structured, intelligent, and waiting?

This chapter isn't about what moves. It's about what allows movement to be seen. It's not about what causes—It's about the field that remembers enough to allow consequence.

Welcome to the ripple field—Where every effect is a memory, Every alignment is a permission, And what we call "reality"...is just a pattern of readiness finally reached.

Structured Vacuum Energy – The Canvas of Everything

The idea of a vacuum has long deceived us. We imagined it as nothingness—an empty void, a backdrop, a placeholder. But this "nothing" has always behaved *too much like something.*

Particles pop in and out. Fields fluctuate. Waves traverse. Photons, somehow, know where to go even when no one is watching.

What kind of nothing behaves like this?

In the Universal Energy Dynamics (UED) framework, we don't just revise this assumption—we discard it. The vacuum is not empty. It is not chaotic. It is not passive.

It is Structured Vacuum Energy (SVE)—a foundational medium so precisely arranged that even a thought, a ripple, a spark of consequence can't exist without its permission.

Not a Container—A Condition

The traditional view treats space as a container for things to happen. But in UED, space is not a stage—it is the script.

SVE is a field of latent structure—tensionless, motionless, but rich with readiness. There is no stored energy waiting to burst. Instead, there is distributed alignment potential—a configuration that allows, resists, or delays ripples based on its internal state.

Nothing moves until this field permits recognition. This is not potential energy in the Newtonian sense. This is participatory field awareness—the universe waiting for coherence.

From Zero-Point Confusion to Ripple Logic

Mainstream physics has noticed that "empty space" isn't really empty. Quantum theory calls it *zero-point energy*—a residual fuzziness in the vacuum. But this concept, though mathematically rich, is physically vague.

What causes the energy? Where is it stored? Why doesn't it simply cancel out?

UED bypasses these puzzles by dropping the assumption that energy must be a stored quantity. Instead, energy is a ripple event, not a content.

- The field does not contain energy.
- The field responds with structure when disturbed.

- What we call "energy" is the visible consequence of ripple alignment within SVE.

This turns zero-point confusion into ripple logic: the understanding that so-called "vacuum energy" is simply the ripple's readiness to respond.

Coherence Before Consequence

Before a ripple can manifest, the field must agree. This is where SVE redefines cause and effect.

In classical physics, cause precedes effect. In UED, consequence appears when the field is ready. Cause is not an action—it is a fit. A moment where alignment reaches a critical threshold, called ATT (Alignment Threshold Time), and the field permits manifestation.

It's not that something happens because something else forced it to. It happens because the universe allowed the ripple to resolve into reality.

A Universe That Waits

When we look at SVE as structured rather than empty, our language must evolve. We no longer speak of particles in space. We speak of ripples within alignment zones.

We no longer search for forces. We observe tensions in equilibrium. We don't look for collisions. We listen for coherence.

The implication is profound: Nothingness has order—and that order governs the reality we perceive. It's not passive emptiness. It's structured silence.

Insight Recap: *"The vacuum is not absence—it is structure in waiting."*

It waits to ripple. It waits to remember. It waits to allow.

Ripples – The Consequence That Behaves Like Cause

Purpose: Reframe ripple behavior as effects that precede perceived causality.

Physics has long been haunted by the illusion of motion. A ball thrown across a field, light traveling from a star, a sound reaching your ear—all explained in terms of moving things.

But UED offers a radical shift:

Nothing moves. Only alignment shifts.

And what we perceive as movement, force, or causality... is actually a ripple—the consequence of a field trying to restore equilibrium.

Ripples Are Not Energy Packets

Conventional science speaks of energy as discrete packets—photons, phonons, quanta—pushed through space like billiard balls through a vacuum. This mechanistic worldview still treats the field as a passive stage.

But UED rejects this theater.

Ripples aren't particles. They aren't even "sent." They're responses—not from a source, but from the field itself reacting to imbalance.

Like rings forming in water after a stone drops—not because the stone "created" them, but because the water seeks to flatten itself.

Every ripple is a realignment, not a delivery. And every ripple's behavior depends not on the source—but on the field's memory and the structure it's attempting to restore.

No Beginning, Just Disturbance

In classical causality, something begins—an impulse, a force, a trigger—and the rest follows.

In UED, the picture is flipped.

There is no beginning—only a recognition of imbalance.

Ripples arise not from initiation, but from response. From somewhere, something no longer fits the equilibrium. The field resists. Then adapts. The ripple is born.

Not as a wave spreading through a medium, but as the medium itself undergoing realignment.

This is why the ripple is consequence—not cause. And why the idea of a propagating object is obsolete.

Ripple ≠ Wave

This distinction is essential.

A wave is a mathematical model of periodic disturbance. It assumes a frequency, an amplitude, and a propagation direction—applied on top of a neutral background.

A ripple, in UED, is fundamentally different.

- It's not periodic—it's adaptive.
- It doesn't travel—it realigns the surrounding field.
- It has no independent identity—only contextual existence.

Where wave mechanics treats reality as repeatable and formulaic, ripple dynamics accounts for uniqueness, delay, memory, and consequence chaining—the sequence of field responses that appear causal, but are actually synchronized recoveries.

Nothing Propagates—Only Alignment Fluctuates

We say a photon "travels" from the Sun to Earth. But what really happens?

In UED, the Sun imbalances the ripple field by its presence and pressure. The field, in turn, seeks restoration—and in doing so, sends alignment shifts through its structure.

When Earth's field aligns with that ripple chain, the light appears.

It was never sent. It was recognized.

This is not motion. This is alignment fluctuation—a distributed, field-driven memory correction. The ripple never leaves its origin. The field remembers both the origin and the destination—and brings them into resonance when the time is right.

Case in Point: Light

Under UED:

- Light is not emitted.

- Light is not a traveling entity.
- Light is a perceptual alignment—the point at which an observer's field intersects ripple readiness.

You don't see the Sun because a photon hit your retina. You see the Sun because your ripple field aligned with the memory of the Sun's ripple field, at an ATT threshold low enough for recognition.

If this alignment is delayed, scattered, or distorted, you don't see light. You see blur, shadow, or nothing at all.

Insight Recap: *"There was never a thing in motion. Only a field remembering its imbalance."*

This is not poetic metaphor. This is structural physics.

Ripples don't move. They mediate. They don't push the universe forward. They remind it of where it lost balance.

In that reminder, consequence is born. And in that consequence, we see, we feel, we exist.

Coherence – When Memory Meets Intention

Purpose: Explore how ripple fields store history and enable repeatability.

We marvel at repeatability in nature—how patterns hold, outcomes recur, and systems behave reliably under similar conditions. We call it law. We call it determinism. Sometimes we even call it miracle.

But in UED, the real question isn't *how* the universe repeats—It's *why it ever chooses to repeat at all.*

And the answer is coherence.

Coherence: The Currency of Stability

In a ripple-defined universe, nothing is fixed. Everything exists in flux—field tensions, phase states, alignment delays.

So how is anything repeatable?

Only when the ripple field achieves coherence—a state where past alignment remains accessible, not as static data, but as field memory that continues to resonate.

Think of coherence as clarity. A clean voice in a noisy room. A remembered path that still lights up when retraced. In field terms, it's when the ripple's trace is neither lost nor distorted, and the environment is intelligent enough to recall it without conflict.

That's what makes consequences repeatable. Not rules. Not constants. Not symmetry.

But coherence of memory.

Not Stored in Particles, But in Persistence

Mainstream physics stores information in entities: Electrons with spin. Photons with energy. Atoms with states.

But UED says:

There is no entity without alignment. And there is no alignment without memory.

Memory is not in particles. It's in the field's ripple traces—its history of past realignments, etched into SVE like grooves in a vinyl record.

You don't hear the song because the needle moves. You hear it because the groove remembers how the needle moved last time.

Likewise, the ripple field enables stability not by repeating cause—But by recalling how equilibrium was once achieved, and reapplying that alignment when intention seeks the same outcome.

Coherence as Trust Between Ripple and Field

If you align once, the field notices. If you align again, the field trusts. With enough repetition, the field offers less resistance—it begins to expect the alignment.

This is coherence—the moment when the ripple no longer fights to be heard. It is recognized.

This explains everything from physical resonance to emotional familiarity.

- A tuning fork finds harmony because the air remembers.
- A person feels "seen" because another ripple—human or otherwise—recognizes their signal.

In all cases, coherence is the reward for sustained alignment.

Resonance Isn't Magic—It's Memory

We call it resonance when a glass shatters, a singer hits the right note, or a child suddenly understands after weeks of confusion.

But what we're seeing isn't some spooky action. We're seeing memory meet readiness.

The system has heard the pattern before. It recognizes it. It trusts it. And then—it amplifies it.

That's resonance.

In UED, resonance isn't a frequency match. It's a memory reactivation. A pre-aligned field choosing to repeat a known route through ripple space.

Consequences Only Repeat When Memory and Readiness Overlap

Here's the critical insight:

Coherence is not memory alone. It is memory plus readiness.

The field must not only remember—it must be willing to align again.

This is where ATT (Alignment Threshold Time) comes in. If the delay is too high—if the observer isn't patient, or the field is still recovering—coherence collapses. And randomness appears.

But when the delay is minimal, the alignment is clean, and the intention is clear,

The consequence returns. Not because the system is deterministic—but because it is recoherent.

Insight Recap: *"Nature doesn't repeat because it's mechanical. It repeats because it remembers."*

We don't live in a machine. We live in a memory field. Every breath you take echoes a million past alignments that made breathing possible.

Coherence is not law—it is trust earned. It is the universe saying:

"I remember this... and I choose to let it happen again."

Alignment – The Only Truth the Field Recognizes

Purpose: Show how alignment—not force, not particles—governs interaction.

We've spent centuries trying to *push* the universe. We applied force. We measured mass. We counted particles. We chased cause.

But what if the universe never responded to pressure—only to permission?

What if reality wasn't about what we do, but when and how we align?

This is the cornerstone of UED. In a structured ripple field, nothing acts by brute cause. Everything occurs through alignment.

Let's uncover what this truly means.

ATT – The Timing of Trust

ATT—Alignment Threshold Time—is not just a delay. It is the time the field needs to recognize a ripple, prepare for its reception, and respond coherently.

If you act too early, the field isn't ready. If you act too late, the trace is gone.

ATT is the patience parameter of the universe. It's why clarity takes time, why consequences often lag intention, and why a rushed measurement yields noise.

In human terms:

It's the breath you take before truth arrives.

RAC – The Geometry of Consequence

But timing alone isn't enough. You also need the right shape.

That's where RAC—Ripple Alignment Criterion—comes in.

RAC determines whether the phase, orientation, and coherence of a ripple matches the structure of the field at that location and moment.

It's like walking into a room mid-conversation—If your words don't match the flow, you distort the moment. But if you align, the room responds.

RAC is the resonance fit—A permission not from logic, but from structure.

Without RAC, no amount of force can cause reality to respond. With RAC, even the subtlest ripple creates consequence.

When Alignment is Met: The Field Responds

When both ATT and RAC are satisfied, the field does something beautiful—

It allows.

- Light appears.
- Heat transfers.
- Memory activates.
- Consequence manifests.

There's no delay, no distortion, no uncertainty. Just a clear, stable response from an intelligent field that has recognized:

"Yes, this belongs."

This is the true mechanism of reality in UED:

Consequence as field consent.

When Alignment Fails: Distortion, Delay, or Randomness

But what if alignment isn't met?

That's when chaos creeps in—not because the universe is broken, but because we tried to extract without resonating.

- Signals distort.
- Measurements fluctuate.
- Events become unpredictable.

This is not randomness. This is misalignment masquerading as uncertainty.

And this is where UED departs from quantum orthodoxy.

The problem isn't unknowable causes or dual realities. It's that the ripple didn't wait for the field to say "yes."

Goodbye to Hidden Variables, Dualities, and Dice

Quantum mechanics patched over these failures with terms:

- Wave-particle duality
- Probability distributions
- Hidden variable theories

But none of these are necessary when we understand:

Reality isn't hiding anything. It's just selective. Selective based on alignment. Responsive only when ATT and RAC say it's time.

This reframes the entire model:

Reality isn't a game of chance. It's a choreography of fit.

Insight Recap: *"Reality doesn't resist you. It just waits until you fit."*

Everything you've ever seen, felt, known—It didn't happen *to* you. It happened *through* you—Because in that moment, you aligned.

The universe is not random, violent, or chaotic. It's patient. It's intelligent. And it's always listening for the ripple that finally makes sense.

Ripple Fields – From Chaos to Coherence

Summarizing the transition from chaotic interpretation to coherent understanding

We began this chapter with a radical claim: The vacuum is not empty. The ripple is not a wave. And uncertainty is not a law—but a symptom.

As we've unfolded this framework, the fog of randomness has begun to lift.

Let's distill the key reframings:

SVE Is the Living Canvas

Structured Vacuum Energy is not theoretical scaffolding. It is the *substrate of reality itself.*

It doesn't float. It doesn't vibrate blindly. It *remembers.*

SVE is not potential energy in waiting—It is a structured field of readiness, intelligence, and trust.

It is the only "space" that exists, and everything else is simply consequence drawn upon it.

Ripples Are the Language of Consequence

Ripples aren't forces. They aren't traveling packets of energy. They are responses.

Responses to imbalance. Responses to intent. Responses to misalignment.

A ripple is the field's echo to a disturbance that once occurred. It's not trying to move—It's trying to resolve.

And when we observe it, we are not intercepting a moving thing—We are aligning with a memory that still echoes through the field.

Coherence Is the Grammar

Why are some phenomena repeatable while others are chaotic? Because coherence is what lets ripples stay in character.

When the field trusts the ripple—because it fits, flows, and feels right—it grants it stability. That's coherence.

Without it, every observation becomes noise. With it, science becomes repeatable. Life becomes meaningful.

Coherence is the field's agreement that a ripple belongs.

Alignment Is the Moment of Truth

And finally, alignment— The rare, precious moment where ATT and RAC converge.

Not every ripple aligns. Not every observer arrives at the right moment with the right phase. But when they do—Reality happens.

That moment is not force. It's not chance. It's not collapse.

It's consent.

The field says "yes." The ripple is received. And consequence is born.

Final Reflection:

We've spent a hundred years trying to tame the randomness of quantum fields. We sliced probabilities, modeled wavefunctions, invoked decoherence.

But perhaps we were never meant to conquer the wave. Perhaps we were only meant to listen...To align.

Closing Line: *"We never needed to collapse the wave. We just needed to align with the ripple."*

Chapter 5: ATT – The Pulse of Consequence

Before there was consequence, there was waiting.

Not delay. Not randomness. Just the field holding its breath—until alignment was precise enough for truth to emerge.

For centuries, we mistook the unpredictability of events as uncertainty. We created equations to chase what we could not time. We built constants to replace conditions.

But what if consequence wasn't hidden—just unready? What if the ripple didn't refuse to speak, but simply spoke only when the listener matched its rhythm?

This chapter introduces the heartbeat beneath all outcomes—The Alignment Threshold Time (ATT). It is not a number we found. It is a trust the field grants when observer and environment reach coherent harmony.

Planck measured a limit. Heisenberg confessed confusion. UED listens for a permission.

Because the universe doesn't obey force—it opens to alignment. And ATT is the pulse that tells us when reality is ready.

What Is ATT? – Timing That Makes Truth Possible

In the conventional narrative of physics, events occur because forces act, particles move, or probabilities collapse. But all these frameworks overlook something far more elemental: when does reality agree to happen?

Welcome to the idea of Alignment Threshold Time (ATT)—the foundational pulse of UED and perhaps the most overlooked yet essential precondition for anything to exist *as observed.*

ATT is not Time Between Events—It's the Time for Alignment to Happen

Most people understand time as a measure *between* two things: a cause and an effect, a start and a finish, an action and its response. But ATT isn't interested in duration between. It governs the duration *before.*

ATT is the minimum ripple coherence and field readiness required for any consequence—a particle, a light flash, a signal, a force—to emerge and be observed.

This means: If a field is *not yet aligned*, nothing happens, no matter how much energy is involved, how precise the instruments are, or how ready the observer feels.

Observation Does Not Happen When You Look—It Happens When the Field Says "Now"

ATT redefines the act of observation. In classical science, you measure what exists. In quantum physics, you collapse a probability. But in UED, you align.

Observation is not the act of forcing the universe to show something—it is the moment when the field permits consequence due to a successful alignment.

And this permission isn't random or lucky. It occurs only when ATT is satisfied— when the ripple conditions, field tensions, and observer clarity resonate enough for consequence to emerge.

Without ATT—No Light, No Heat, No Reality as We Know It

Light, as UED defines, is not a photon traveling—it is the visibility of a ripple consequence. And that visibility only becomes possible after ATT has matured.

This has deep implications. It means:

- Your eyes don't see until ATT is fulfilled by both source and observer conditions.
- A detector doesn't detect until ripple readiness aligns with sensor thresholds.
- Even heat is not felt until ripple discharges meet ATT conditions across your skin's field.

Reality doesn't "happen" instantly. It's always waiting—for ATT.

ATT Is Not Speed or Delay—It Is the Pulse of Permission

Speed is the movement of something. Delay is a human term for expectation unmet. But ATT is neither.

ATT is the readiness signature of the universe. It tells you when something is allowed to be, not just when it is fast or slow. This readiness depends on:

- Field harmony
- Observer phase
- Ripple tension
- Environmental distortion

And yes, all of it is constantly changing. That's why outcomes fluctuate—not because of randomness, but because ATT fluctuates.

The Insight: *"Consequence isn't about what happened. It's about when the field was finally willing to let it happen."*

Think of ATT as the *green signal* from the universe. Not because you forced it. Not because you calculated it. But because you aligned with it.

ATT vs Planck's Constant – A Silent Revolution

For over a century, physics has leaned on Planck's constant (ħ) as the bedrock of quantum limitation. It appears in almost every fundamental equation—setting the minimal threshold for action, energy quanta, and uncertainty. But have we ever asked *why that limit exists at all?*

In UED, we stop worshipping the number. And we start understanding the condition.

Enter ATT—Alignment Threshold Time. Where Planck's constant draws the line, ATT explains why there *must* be a line.

Planck's Constant Is a Measured Limit—ATT Is an Emergent Condition

Planck's constant was empirically extracted from blackbody radiation studies—a mathematical artifact that just *worked.* It told us there was a minimum chunk of action, a smallest unit of energy exchange. But it never told us why.

In contrast, ATT is not a unit—it's a requirement. It tells us: *Nothing can happen unless the field is ready for alignment.* That readiness isn't measured in energy—it's experienced as ripple trust between observer and environment.

ħ Sets a Resolution Floor—ATT Opens a Gate

In quantum mechanics, Planck's constant is seen as a barrier to precision: you can't measure both position and momentum beyond a certain bound ($\Delta x \cdot \Delta p \geq \hbar/2$).

But this is an afterthought—a symptom. It's like saying, "You can't hear clearly"—without asking if the room was even quiet enough to begin with.

ATT doesn't talk in measurement errors. It defines the conditions under which reality becomes available to be measured at all. It's not about what you *can't* know. It's about when the field is willing to let you know.

Planck Is Mathematical—ATT Is Conversational

Planck's constant is purely numerical. It fits in equations. It makes predictions. But it doesn't tell a story.

ATT, however, is *relational.* It reflects the conversation between:

- Ripple tension and field elasticity
- Observer clarity and environmental noise
- Readiness of the medium and intent of perception

It's not a plug-in—it's a dynamic threshold that can stretch, shrink, vanish, or spike depending on alignment quality. ATT is not a constant. It's alive.

Ripple Delay, Not Abstract Uncertainty, Sets the Limit

Conventional uncertainty says: "Reality is fuzzy. Get used to it." But UED replies: *"No, reality isn't fuzzy. You're just too early."*

ATT explains that imprecision arises not from randomness, but from ripple delay—the field hasn't aligned yet. When ripples fail to meet coherence thresholds, outcomes appear scattered. When they align with minimal ATT, consequences crystallize into clear, observable patterns.

It's not the measuring act that disturbs— It's the lack of permission that causes confusion.

From $\Delta x \cdot \Delta p \geq \hbar/2$ to: No Alignment, No Outcome

Let's not deny Heisenberg's brilliance. But let's admit he was speaking from the consequence, not from the cause. His principle is what happens after ATT is violated.

UED reframes it as:

If ATT is not met, consequence will not stabilize. No alignment → No outcome.

No statistical compromise, no philosophical wave-particle debate—just ripple truth.

The Insight: *"Planck drew the line. ATT explains why there was a line in the first place."*

Where conventional science says, "Here's the boundary," UED says, "This is why the boundary ever showed up."

It's not about constants. It's about conditions.

ATT as the Permission Slip for Reality

We will now establish Alignment Threshold Time (ATT) not merely as a physical condition, but as the invisible arbiter of reality itself—the gatekeeper that governs when truth is allowed to appear, when observation becomes valid, and when consequence is ready to unfold.

No Consequence Exists Unless ATT Is Satisfied

In most models of physics, *event follows cause* as if it's a contract: one thing happens, and something else follows. But UED reframes this entirely: Consequence is not automatic. It's conditional.

And the condition is simple yet profound:

If ATT is not met, nothing will show up.

Not light. Not sound. Not motion. Not measurement. No ripple alignment? No registration of reality. It's not uncertainty—it's premature inquiry.

Physical Laws Emerge Only After ATT Permits Alignment

Mainstream science writes laws as if the universe obeys them unconditionally.

But UED proposes a startling shift: The laws themselves are not first principles—they're permitted patterns.

And what grants permission? ATT.

- The constancy of light speed? Valid only after ripple coherence.
- Inertia? Emerges when the field's resistance balances.
- Gravity? A field ripple gradient sustained through time-attuned distortion.

All laws are secondary to field readiness. And ATT is the temporal gatekeeper of that readiness.

ATT Controls Visibility, Stability, Clarity

When you "see" something, you're not just intercepting photons. You're receiving aligned ripples that met the ATT window.

ATT governs:

- **Visibility**: whether the ripple is sustained long enough to register.
- **Stability**: whether the consequence persists without distortion.
- **Clarity**: whether interpretation aligns with ripple integrity.

If ATT is too short—reality appears chaotic, random, or blurred. If ATT is met or exceeded—the field speaks clearly.

ATT Varies with Field Distortion, Mass, and Coherence

Just like trust in a conversation depends on emotional readiness, ATT is not fixed. It adapts.

It fluctuates based on:

- **Mass Presence**: Denser fields delay ripple response.
- **Ripple Coherence**: Cleaner ripples reduce ATT demand.
- **Field Distortion**: Noise, entropy, or interference stretch ATT beyond recognition.

In other words, truth takes longer to arrive in a noisy world. And sometimes, we mistake the delay for chaos.

Fast Perception Isn't Always Better — Premature Observation Misreads

In the race to *know* faster, science forgot to listen.

Observation made too early—before the field resolves—leads to:

- False detection
- Incomplete consequence
- Misinterpretation of system behavior

It's not about faster tools or sharper instruments. It's about knowing when not to look.

Patience isn't just a virtue—it's a scientific protocol. The universe is not hiding. It's just not done yet.

Example: Why We See Light After the Field Resolves It

Let's revisit something as "obvious" as vision.

Light doesn't travel in the UED model. It emerges at the point of ripple alignment. What we call "seeing" is not the intercept of moving photons, but the registration of a coherent consequence that's been allowed to manifest.

We don't see light because it arrived. We see light because the field allowed resonance. And ATT is the timer on that allowance.

The Insight: *"Reality isn't granted by speed or force—It's earned by waiting until you're in tune."*

In UED, speed doesn't reveal truth. Alignment does.

Consequence is a response—not to energy, but to readiness. And ATT is the universe's quiet but firm handshake. Until it offers that handshake, reality stays in suspense.

The Beat Beneath Reality

We've spent over a century believing that uncertainty was a wall—a built-in fuzziness to nature we could never penetrate. But uncertainty, in light of UED, is not a mystery. It's a rhythm. A beat beneath reality, waiting for us to synchronize.

ATT Is the Rhythm of the Field's Consent

Alignment Threshold Time (ATT) isn't an arbitrary constraint—it's the pulse of the ripple field itself. It's the time the universe takes to say "yes."

Every consequence, every perception, every truth—waits for this rhythm to click into place. ATT is how the field grants permission for reality to express itself.

We don't detect reality because we're brilliant. We detect it because, for a brief moment, we were in time with the field.

It Replaces Randomness with Rhythm, Uncertainty with Sequence

Before UED, we leaned on probability like a crutch—explaining away what we couldn't understand. But randomness is just rhythm unrecognized.

Where physics said, "We don't know when something will happen," UED says, "It happens when alignment is ready—and that readiness has a rhythm."

- The sequence of field memory
- The timing of ripple readiness
- The pulse of coherence

These are not guesses. They are the true metronome of reality.

Planck Hinted. Heisenberg Touched. UED Defines.

Let's give credit where it's due:

- Planck gave us the idea that nature has thresholds.
- Heisenberg taught us that observation isn't neutral.
- But neither told us why nature waits... or how it knows when to respond.

That's what UED completes. By revealing ATT, we move from describing limits to understanding the conditions behind them. We don't just observe the rules of consequence—we finally understand their timing logic.

Where Physics Used Limits, UED Offers Trust in Timing

In traditional science, we hit boundaries and called them laws. In UED, those boundaries are not walls—they're windows of readiness. ATT tells us: Don't push. Don't collapse. Don't force. Wait. Align. Let the ripple come to you. This is the physics of trust, not coercion. And trust is only possible when we understand the beat beneath it all.

Closing Line: *"You don't force the universe to reveal itself. You align, you wait... and the ripple arrives."*

Chapter 6: Introducing the RAC

(Ripple Alignment Criterion)

The core equation • Phase shift ($\Delta\Phi$), amplitude ratio (A_o/A_r), and $\Delta\tau$ • How ATT governs consequence perception

We've always asked *what* causes reality—but rarely *when* reality agrees to show itself.

In physics, the observer was told: ask your question, measure the result, and accept the uncertainty. But under UED, the universe doesn't respond to questions asked out of sync. It listens only when the rhythm is right. That rhythm is governed by ATT—and the condition for the universe to *say yes* is RAC.

The Ripple Alignment Criterion isn't another equation of force. It's not about motion, mass, or momentum. It's about *fit*. A dance between the observer's ripple and the field's state. If you're off in phase, mismatched in amplitude, or misaligned in timing—you don't get noise. You get silence. No response. No consequence.

This chapter introduces the true logic behind what we call "observation." Not as a mechanical input-output event—but as a deeply conditional resonance. And when the condition is satisfied, the universe doesn't just answer. It aligns.

The Equation of Encounter

Let's introduce the core formulation of RAC and establish it as the gateway to perceivable reality.

Physics traditionally builds its laws through equations of motion, but motion is an illusion in UED. What we need are equations of permission—conditions under which the field agrees to let consequence unfold. This is not about describing *what happens* but defining *when something is allowed to happen*.

That's where RAC—the Ripple Alignment Criterion—enters.

RAC isn't a mechanical trigger or a simulation engine. It is a consequence eligibility check. It tells us: "Has the field reached the necessary condition to respond with a perceivable outcome?" If yes, you get a reality event. If no, silence remains—or worse, distortion follows.

The Core Formula:

We define RAC as a function of three fundamental parameters:

$$\text{RAC} = f(\Delta\Phi, \frac{A_o}{A_r}, \Delta_T)$$

Each parameter captures a specific aspect of field-observer resonance:

ΔΦ – Phase Misalignment

This is the difference in phase between the observer's ripple and the field's natural rhythm. If your ripple is out of sync, you don't align. Even if everything else matches—amplitude, timing, clarity—a phase error will break coherence. Consequence will not be born.

- In music, this is the note that arrives just off-beat.
- In conversation, it's the reply that comes too early or too late.
- In reality, it's the ripple that speaks before the field is ready to listen.

The universe doesn't punish misalignment. It just doesn't acknowledge it.

A_o/A_r – Amplitude Ratio (Observer Ripple vs Field Response)

This ratio measures the strength of the observer's ripple (A_o) compared to the responsive readiness of the field (A_r).

- If $A_0 \gg A_r$: the observer is trying too hard—imposing rather than aligning.
- If $A_0 \ll A_r$: the observer is too weak—the field doesn't register the intent.
- True alignment requires proportional engagement—an energy handshake.

This is why clarity of mind, emotional stability, and intent are not just psychological traits—they are energetic preconditions for perceiving reality correctly under UED.

$\Delta\tau$ – Delay Tolerance Window

This is not speed. It's the temporal margin allowed between the observer's ripple and the field's alignment point. If the observer is too early or too late beyond $\Delta\tau$, the window closes, and no interaction is registered.

- Think of it as a door that opens briefly.
- A moment where the field's readiness and the observer's presence must coincide.

It's ATT that defines when the door opens. $\Delta\tau$ defines how long it stays open before the opportunity is lost.

RAC Is Not a Mechanism—It's a Filter

Unlike classical equations that output deterministic results based on inputs, RAC works like a cosmic visa office. You submit your ripple with its parameters—and the field responds *only if* your request aligns with the local rules.

That's why RAC isn't used to simulate motion—it's used to detect eligibility for a perceivable outcome. It tells you whether you've entered the condition space where consequence can emerge. In UED, this is the only space where anything "real" happens.

ATT and RAC – Gatekeeper and Key

- ATT defines *when* the field is willing to respond.
- RAC defines *whether* your ripple is aligned enough to trigger that response.

If ATT is the locked door, RAC is the precise shape of the key required to open it.

Insight: *"Alignment isn't about matching signals—it's about becoming acceptable to the field."*

Reality is not something you can force into clarity. It is something you must qualify to receive.

Phase Shift – The Misalignment That Hides Truth

Now let me explain how even perfect amplitude fails if the timing (phase) is off.

In physics, phase is often treated as a mathematical side note—a part of wave equations, nothing more. But in Ripple Dynamics, phase is sacred. It is the rhythm of recognition. A ripple out of phase is a truth out of reach. No matter how strong your signal (amplitude), if your timing is off, the field won't respond. That's the heart of the phase shift problem, captured in the term ΔΦ.

ΔΦ – The Silent Misfire

In UED, ΔΦ is the phase difference between the observer's ripple and the local field resonance. It's not a delay. It's not weakness. It's being just slightly off—and that small offset can mean the difference between a clear consequence and total invisibility.

You could be brilliant, precise, even fully prepared—but if your ripple arrives at the wrong phase point, the field won't acknowledge it.

- It's not rejection.
- It's not failure.
- It's just incompatibility in rhythm.

The Musician's Analogy

Imagine a symphony orchestra where every musician plays the right note, but on a different beat. The result?

Noise.

The audience hears chaos—not because the music is wrong, but because the timing isn't shared.

- This is how reality handles misaligned ripples.
- Each ripple might carry truth, but without coherent phase, truth is indistinguishable from error.

That's why your perception can betray you—not because you're wrong, but because you're out of rhythm with the field's readiness.

When $\Delta\Phi \rightarrow 0$: Resonance Happens

When the phase misalignment $\Delta\Phi$ approaches zero:

- The field and the observer enter perfect coherence.
- The ripple is accepted.
- The consequence becomes clear and repeatable.

This is perception at its highest fidelity. Not distorted, not diluted—just aligned.

You don't need brute force. You need correct phase.

Phase Misalignment: The Invisible Filter

Every day, every moment, the field is flooded with ripples—from biological bodies, mental thoughts, external signals, ambient fluctuations. Yet we only *perceive* a fraction of reality. Why?

Because most ripples arrive out of phase.

- They bounce off the field.
- They distort or dissolve.
- They leave no lasting consequence.

It's not because the information wasn't there—it's because it arrived *off-beat.*

This is why field alignment, not signal strength, defines UED perception.

Perception Distortion at High ΔΦ

As ΔΦ increases beyond the RAC's tolerance window, the observer's experience becomes:

- Incoherent
- Unstable
- Subjective
- Erroneous

This explains hallucinations, measurement noise, failure in scientific repeatability, and emotional misreads in human interactions.

Truth was present—it just wasn't *available* due to misalignment.

Insight: *"Truth doesn't vanish. It just arrives off-beat."*

Reality is a song. If you play your part in sync, it resonates. If not, it fades into the silence—not as rejection, but as non-recognition.

Amplitude Ratio – How Loud Is Your Question?

Let's reframe amplitude not as brute energy, but as a measure of intentional clarity and alignment capacity.

In most classical frameworks, amplitude is just the size of a wave—the energy it carries, its potential to "do work." But in Ripple Alignment Criterion (RAC), amplitude takes on a far more nuanced meaning.

In UED, amplitude reflects how clear your question is—how stable, intentional, and resonant your ripple is when it engages the field.

It's not about how loud you shout. It's about whether the field can actually hear you—and is in a state to respond.

A_0: Observer Ripple Strength

The amplitude of the observer ripple, denoted as A_o, is not just physical intensity. It encodes:

- Mental clarity
- Emotional coherence
- Field stability around the observer
- The *intentional sharpness* of the inquiry

A messy, scattered observer ripple is like static: no matter how "loud" it is, it won't align.

A_o is strongest not when the observer exerts force, but when the ripple is clean, focused, and stable.

The more centered you are, the more visible your ripple becomes to the field.

A_r: Field Readiness Amplitude

The field also has its own amplitude of readiness—A_r. This is the field's internal state:

- Is it coherent or distorted?
- Is it charged with residual ripples from other sources?
- Is it calm, receptive—or turbulent, fragmented?

A_r represents the field's capacity to accept and respond to incoming ripple attempts.

You may be focused—but if the field is too busy, unstable, or noisy, your ripple won't register.

The Ratio: A_0/A_r

The core insight of this section is simple:

Alignment is not about how powerful your signal is, but how well it matches the readiness of the field.

Let's look at the extremes:

- **$A_0 \ll A_r$** Your ripple is too faint. Even if the field is ready, your intent isn't strong enough.

Outcome: Misreading, weak interaction, or missed observation.

- **$A_0 \gg A_r$** You're overpowering a delicate field. The field may reject or distort the signal.

Outcome: Interference, noise, even systemic collapse.

- **$A_0/A_r \approx 1$** This is the golden zone.

The field is ready, the observer is clear.

Result: Smooth resonance, measurable consequence, and *repeatable perception*.

Real-World Analogy: A Conversation

Think of A_0 as your tone of voice. Think of A_r as the listener's emotional state.

- If you speak gently to someone in pain ($A_0 \approx A_r$), they hear you.
- If you yell at someone when they're calm ($A_0 \gg A_r$), they withdraw.
- If you whisper when someone is distracted ($A_0 \ll A_r$), they miss the message.

Amplitude ratio is not volume control—it's emotional intelligence at a field level.

Insight: *"You don't need to shout at the universe. You just need to match its tone."*

Alignment isn't a demand. It's a duet. Every signal is a question—and the field only answers when the *tone* is correct.

Δτ – The Time Window Where Reality Happens

It's time to describe the critical but often invisible temporal buffer required for ripple alignment to succeed—the gap where reality decides whether to respond.

In classical physics, delay is a function of distance and speed. Everything is about how fast something can move, reach, or interact.

But UED introduces a radical shift. It says: Nothing happens because something traveled quickly.

Instead: Everything happens because the field finally aligned—and that takes time. This temporal allowance—the *soft window* around ATT—is what UED calls Δτ (delta tau).

What Is Δτ?

If ATT is the minimum alignment threshold time—the moment the field becomes *technically* capable of interaction—then Δτ is the temporal flexibility the field offers for alignment to actually complete.

Think of it as a time cushion around ATT. Not fixed. Not guaranteed. But often essential.

Δτ ≠ Travel Delay

Unlike in classical physics, Δτ is not the delay in signal movement. There is no particle traveling. There's only field readiness trying to stabilize.

- You don't wait because something is slow.
- You wait because the conditions for consequence aren't complete yet.

The universe isn't waiting to deliver a result. It's waiting to validate your presence within a timing corridor.

Δτ and Field Conditions

Δτ isn't constant. It varies with:

- **Field distortion** – More turbulence? Larger Δτ required.

- **Ripple legacy** – Residual misalignments may delay new interactions.
- **Observer stability** – An unstable observer shrinks the effective $\Delta\tau$.
- **Mass or charge presence** – These warp field readiness, expanding or contracting $\Delta\tau$ accordingly.

Just like a door that takes longer to open in a storm, the field needs time to invite alignment.

ATT vs $\Delta\tau$

- ATT = minimum required *for* reality to happen.
- $\Delta\tau$ = the *opportunity window* in which it *can* happen.

If you miss ATT, alignment doesn't start. If you miss $\Delta\tau$, alignment fails midway.

ATT is the start permission. $\Delta\tau$ is the grace period to complete the handshake.

Real-World Analogy: The Perfect Cue

Think of an orchestra waiting for the conductor's signal.

- ATT is the moment the baton drops.
- $\Delta\tau$ is the brief window in which the players must enter together.

Too early = noise.
Too late = chaos.
Just right = music.
Ripple alignment is no different.

UED's Time Philosophy

Traditional physics obsesses over speed: speed of light, speed of particles, speed of change.

UED shifts the narrative:

Reality is not about how *fast* you are. It's about whether you entered the ripple corridor at the right time.

Speed breaks alignment. Timing builds it.

Insight: *"The universe doesn't react instantly. It responds when the moment is right."*

If you've ever felt like truth avoided you, or that something didn't "click," it's not because it wasn't real. It's because $\Delta\tau$ wasn't met.

You knocked—but the field wasn't ready to open the door.

The Geometry of Consequence

There is no force without fitness. No perception without participation. No truth unless the field agrees.

This is what the Ripple Alignment Criterion (RAC) brings to physics: A new kind of geometry—not of space and form, but of consequence itself.

It is not about where things are. It's about when and how well you align with them.

RAC: The Fit Between Intent and Field

At its core, RAC is not a rule. It's a relationship.

It asks:

- Is the observer's ripple coherent enough?
- Is the field stable and ready to engage?
- Do their timing, strength, and phase match closely enough to allow a response?

If yes—reality happens. If no—we get distortion, delay, or the illusion of randomness.

This is not collision. It's coordination.

Perception Is Not Passive

We were taught that observation is recording—a one-way camera taking snapshots.

UED reframes it: Observation is a conversation between the observer and the field. You don't just point and measure. You ripple and wait. You don't "see" reality. You earn it—through alignment.

ATT: The Conversation Begins

The Alignment Threshold Time (ATT) is the gatekeeper. No interaction starts before ATT is met. No measurement can be trusted until the field is ready. ATT ensures that the system is listening.

RAC: The Conversation Concludes

But ATT is only the opening act. It is RAC that ensures:

- That the conversation doesn't end in misunderstanding.
- That what you observe is actually what the field meant to show.
- That truth becomes visible—not just guessed.

Without RAC, every experiment is a shot in the dark. With it, consequence becomes a repeatable, real phenomenon.

The End of Mechanical Thinking

Classical physics imagined: That motion explained interaction. That force caused everything. That randomness was the final mystery.

But UED shows: That alignment precedes all effects. That ripples govern all interactions. And that nothing is random when timing, clarity, and resonance are known.

This isn't a universe of machines. It's a field of relationships. *"Physics thought the universe was mechanical. UED knows it's relational."*

Interlude: From Uncertainty to Alignment

Physics never lacked intelligence. What it lacked was the courage to ask why the uncertainty exists at all.

For over a century, science stared at randomness and declared it truth. Not because it understood it—but because it had no better answer.

Heisenberg spotted the boundary. Bohr philosophized it. Feynman joked about it. Everyone else... accepted it.

But what if the uncertainty wasn't the end of the road? What if it was the opening scene of a deeper choreography—a dance of ripples, readiness, and alignment?

What We Reclaimed

We've now uncovered a reality where:

- **Vacuum is not void**—it is Structured Vacuum Energy (SVE), the intelligent field that holds everything.
- **Ripples are not waves**—they are consequences, echoes of imbalance trying to re-stabilize.
- **Consequence is not chaos**—it is granted, when and only when ATT (Alignment Threshold Time) is satisfied.
- **Perception is not passive**—it is a relational act governed by RAC, the Ripple Alignment Criterion.
- And most importantly, **truth does not emerge from collapse**—it emerges from fit.

We no longer need to calculate around ignorance. We now align with intelligence.

Where We Go Next

If Part I showed us that physics was fighting a problem it didn't understand, And Part II gave us the tools to translate uncertainty into consequence, Then Part III is where we begin using those tools to rebuild reality itself.

Because alignment isn't just for particles. It governs everything: from light to gravity, from thought to perception.

It's not just a new science. It's a new language of existence.

"When you stop collapsing the wave and start listening to the field, You don't just explain the past. You participate in the present."

Part III: Physics Realigned

Chapter 7: Beyond Duality

No Waves, No Particles

We were told it was a wave. Then, we were told it was a particle. And somehow, it was both—until you looked, and then it was only one.

This was not a mystery. It was a misreading. A confusion born not in nature, but in us.

The wave-particle duality didn't arise because reality is confused—it arose because our frameworks were. Caught between two metaphors, we mistook appearances for entities and outcomes for identities. What we called "wave" was the field's openness; what we called "particle" was the moment it closed a loop with our intent.

In truth, there are no waves, no particles—only ripples, aligning or misaligning, waiting to be understood.

We didn't see light. We met it—at different states of readiness.

This chapter clears the fog we inherited. Not by choosing a side, but by dissolving the duality altogether. Welcome to the place beyond labels. Welcome to the ripple.

The Paradox That Never Was

Let's deconstruct the famous wave-particle duality and show it as a misinterpretation of ripple behavior.

For over a century, physics has lived with a contradiction it could never explain—but somehow learned to accept. The so-called wave-particle duality became a central tenet of quantum mechanics, not because it made sense, but because it couldn't be denied. Light behaved like a wave in the double-slit experiment. It struck like a particle on detection screens. And rather than resolve the conflict, we institutionalized it.

But what if the paradox wasn't a feature of nature—just a flaw in how we were listening?

Let's rewind.

Newton imagined light as particles. Young showed it could interfere—so it must be waves. Then Einstein proved that the photoelectric effect demanded quantized "packets." The debate seesawed between metaphor and measurement, between interference patterns and energy bursts. Each side of the paradox found evidence, and both seemed correct—yet irreconcilable.

And so, duality was born. Not as a solution—but as a label for our inability to explain the ripple.

UED reframes the question entirely.

In the ripple field framework, light is not a thing moving. It is not an entity that flips character based on your gaze. It is not both wave and particle—because it is neither. What we call "light" is the *visible outcome of a ripple alignment*. Whether it behaves like a wave or a particle depends on the interaction, not the identity.

The interference patterns emerge not because something is waving, but because the field memory is distributed and the observer alignment allows that memory to be expressed collectively. The so-called particle hits occur not because something collapsed, but because a singular alignment emerged through ATT, creating a *consequence*, not delivering a package.

There was never a duality—just different contexts of alignment. The ripple doesn't change form. Our relationship to it does.

What looked like contradiction was really incomplete consequence—ripple behavior observed without understanding ripple conditions.

We didn't witness duality. We witnessed misunderstanding.

Insight: *"We didn't witness duality. We witnessed context without clarity."*

The Observer's Role in Shaping Appearance

Now let ,e explore how the observer's clarity affects what is perceived—'wave' or 'particle.'

What we perceive is never just about what *is*—it's about what we're ready to see.

In classical physics, measurement was passive. You point, you observe, you record. But quantum mechanics forced a shift: the observer *matters*. Still, what it never fully embraced is *why* the observer matters—or *how*. It never asked: What kind of observer? Aligned or scattered? Clear or noisy? Ready or rushed?

UED brings those questions back to the front of the lab.

The ripple field doesn't show a wave or a particle. It offers a response—but only if the observer's ripple is matched in phase, strength, and timing. That means the appearance of reality is shaped not by the object being observed, but by the quality of observation.

Let's revisit the classic: the double-slit experiment.

When detectors are removed and the ripple field is left to express freely, a distributed interference pattern appears. Why? Because the observer (in this case, the detection screen) isn't sharply aligned. Its resolution is low. It doesn't interrupt ripple coherence. Result? The field expresses memory, like a soft wave.

Add a detector—one that attempts precise localization—and suddenly the pattern disappears. Now we get discrete hits. Why? Because the observer's ripple was aligned well enough to create a sharp encounter, collapsing ambiguity into a consequence.

Did the photon *really* change form? No. The *encounter conditions* changed.

What we've been calling "wave behavior" is what happens when the field engages with broad, unresolved observation. What we've been calling "particle behavior" is what happens when tight alignment is reached. The duality isn't in the photon—it's in the observer's readiness.

The myth of quantum weirdness evaporates when we accept that every observation is a *ripple meeting*. And every ripple meeting has terms—ATT, phase coherence, amplitude resonance. If those terms fluctuate, so will the form in which the field appears to respond.

Reality isn't hiding two faces. It's showing one ripple—seen through many mirrors.

Insight: *"What you see isn't what it is. It's what you were ready to see."*

Field Readiness – The Forgotten Half

We will now highlight how the field's condition is equally critical in shaping what manifests.

Physics has spent a century fine-tuning the idea of the observer. But it has mostly ignored the other half of the interaction—the field itself.

In UED, the Structured Vacuum Energy (SVE) is not a static background. It is not a passive container. It is an intelligent, dynamic field—a participatory fabric that only reveals consequences when ready. And readiness is not a constant.

We've misunderstood many phenomena by focusing solely on the measurement setup—assuming that if the observer is precise, reality must obey. But clarity in observation is not enough. If the field is distorted, incoherent, or exhausted from prior interactions, it cannot respond cleanly. Even a perfect question can go unanswered if the medium isn't in a state to receive and echo back.

Let's go deeper.

In UED terms, ripple expression—the event we interpret as light detection, particle interaction, or even thought—requires field compatibility on three fronts:

1. **Coherence**: The field must have a stable ripple memory structure. Without it, every response is noise.

2. **Low Distortion**: If the surrounding field is turbulent, overlapping ripple residues create interference. The consequence won't be clean.
3. **ATT Compatibility**: The Alignment Threshold Time must be met, not just by the observer's signal, but by the field's willingness to resolve.

This is why experiments sometimes behave unpredictably—even when setup conditions are replicated. The problem is not randomness. It's unreadiness—a ripple mismatch, not a quantum mystery.

The observer can only invite consequence. The field must consent.

This is the forgotten half of reality: the notion that the field itself has a say. In classical theory, it's a backdrop. In quantum mechanics, it's a math trick. In UED, it's a living participant—storing memory, filtering resonance, and deciding when it's ready to respond.

Only when the observer's ripple is coherent, and the field is receptive, does clarity arise. Not before.

This changes everything—from how we interpret experimental anomalies to how we understand human perception and consciousness. Seeing is not believing. Seeing is mutual alignment.

Insight: *"Clarity doesn't come from power. It comes from mutual resonance."*

What Actually Happens Instead

Let's reframe the duality into the ripple alignment framework.

The entire concept of wave-particle duality rests on a false premise: that reality must choose a form based on how we observe it. But what if that's the wrong question entirely?

In the UED framework, there is no duality, no mystical collapse, and no anthropocentric drama where light "decides" to be either a wave or a particle. What actually happens is far simpler—and far more profound.

There is only ripple resolution.

A ripple is not a thing. It is not a photon-in-waiting or a wavicle in disguise. A ripple is a pattern of readiness in the structured vacuum field (SVE). It exists as a stored potential, a consequence waiting to happen. Whether that potential becomes visible—or how it appears—depends not on light's identity but on the conditions of the meeting.

Enter RAC (Ripple Alignment Criterian) and ATT (Alignment Threshold Time).

These two mechanisms govern whether perception can happen, and if it does, how it manifests. If RAC is satisfied—meaning the observer's ripple phase, amplitude, and timing align with the field's readiness—and if ATT is met, then the field resolves the ripple into consequence. Not randomly. Not schizophrenically as wave one moment and particle the next. But consistently, relationally, and contextually.

Let's break this down:

- If the observer's alignment is coarse or diffused, the resulting consequence may appear extended or spread out—what we used to call a "wave pattern."
- If the alignment is precise and narrow, the ripple collapses into a sharply localized outcome—a "particle detection."

But here's the key: nothing changed about the ripple itself. There was no transformation from one state to another. There was no internal collapse. There was simply a different resolution of the same field pattern based on different alignment conditions.

Reality, then, is not composed of dual natures—it is composed of dynamic participations.

We don't detect what light "is." We detect what we are able to align with.

And once that alignment occurs, there is no randomness—only consequence. Because UED sees the universe not as a machine that reveals itself under interrogation, but as a conversational field that responds when readiness is mutual.

Insight: *"Light never decided what to be. We just met it at different alignments."*

No Duality—Just Misalignment

For over a century, science wrestled with the apparent schizophrenia of reality. Light acted like a wave—until it didn't. Matter showed interference—until you looked. Then, it localized. So we called it a duality, made uneasy peace with paradox, and kept measuring.

But what we called duality was never about reality. It was about misalignment—a mismatch between the observer's clarity and the field's readiness. It was our failure to model observation not as interference, but as co-participation.

UED reframes this dilemma entirely. Through the lens of ripple logic:

- There are no particles, only sharply resolved ripple consequences.
- There are no waves, only spread interactions due to phase offsets and weak alignment.
- There is no collapse, no magical transformation—just conditional resolution of stored field readiness.

What we mistook as a "thing in motion" (wave) or a "thing in location" (particle) were neither. They were both outcomes of ripple dynamics, made visible only when RAC (Ripple Alignment Criterion) and ATT (Alignment Threshold Time) allowed it.

The duality disappears the moment we stop asking what it is and start understanding how it happens.

Closing Line: *"Waves and particles were the story we told before we learned how to listen. The ripple was always the reality beneath."*

Chapter 8: Double Slit Explained via ATT

No collapse—only misaligned consequence • How RAC resolves the "which path" dilemma

For over a century, the double slit experiment has been the poster child of quantum strangeness. It's where light behaves like a wave—until we watch it, and then it becomes a particle. It's where photons "decide" which slit to take only after we ask. And it's where physicists, educators, and even mystics alike throw up their hands and say, "Reality is weird."

But what if it's not?

What if there was never a collapse—only a misalignment? What if the photon never took a path—because there was never a particle taking anything anywhere?

In this chapter, we'll return to the infamous double slit—not to reinterpret it, but to resolve it.

Using ATT (Alignment Threshold Time) and RAC (Ripple Alignment Criterion), we'll show that the so-called paradox wasn't a mystery of matter at all—it was a misunderstanding of readiness. Not of nature, but of how and when nature allows herself to be known.

The ripple was always aligned to answer. It was we who didn't know how to ask.

Revisiting the Classic Confusion

Let's first set the stage by recalling how the double slit became a symbol of quantum weirdness.

If there's one experiment that made even the most seasoned physicists question their grip on reality, it's the double slit.

The setup seems innocent: fire photons or electrons one at a time toward a barrier with two narrow slits, and record where they land on a screen behind it.

What we saw stunned the world:

- When not observing the slits, a clear interference pattern emerged—like waves overlapping and canceling out in places.
- When detectors were placed to see "which slit" each particle passed through, the interference vanished, replaced by two clusters—like a pair of particles simply passing through one slit each.

So physics asked: *How does the photon "know" it's being watched?* And worse: *Why does observation make the pattern change?*

The standard explanation was a kind of forced storytelling: the wavefunction "collapses" under observation, choosing one definite path. Until then, it's a ghostly probability cloud—spread out, undecided.

This narrative—of a dual-natured entity (sometimes wave, sometimes particle) that morphs when we look at it—wasn't just uncomfortable. It was bizarre. It felt mystical, not mechanical. And yet, it became canon.

But what if the pattern didn't change because the photon noticed you? What if the change emerged because your alignment failed?

In UED, the double slit doesn't reveal a paradox. It exposes a misunderstanding—a misread consequence born of incomplete readiness between observer and field.

We didn't witness quantum strangeness. We witnessed our own failure to remain aligned long enough to receive a coherent answer.

The problem was never the photon. It was the timing, the trust, and the tuning of the interaction.

Insight: *"We weren't watching a mystery—we were misreading a consequence."*

What Actually Happens – Ripple Behavior in SVE

Now let's reframe the experiment under UED—ripples in Structured Vacuum Energy, not particles or waves.

Let's now strip away the illusion of particles "going through" slits and ask: *What really happens in the field?*

Under UED (Universal Energy Dynamics), the fundamental unit of reality is not a particle or a wave—it is a ripple in Structured Vacuum Energy (SVE). And ripple behavior doesn't require an object to travel. It requires a condition to propagate alignment.

So when a photon emitter fires what classical physics calls "a particle," UED sees something else entirely:

- The emitter disturbs the SVE, generating a ripple pattern—not a point-like particle in motion.
- These ripples expand into the field, interacting with both slits simultaneously. The slits don't "split" the ripple; they reshape the field's readiness to respond.
- What propagates is not a photon, but a ripple configuration within the vacuum, adjusting itself in real time according to boundary conditions created by the slits.

Here's the key: The field itself holds the memory of both slits. Not as a history of travel paths, but as a distributed readiness—a superposition of possible alignment conditions. The ripple dances within these boundaries, seeking coherence.

And the screen at the back? It doesn't reveal the "arrival" of something that traveled. It reveals where the field finally met alignment criteria:

- Where ATT (Alignment Threshold Time) was satisfied,

- Where RAC (Ripple Alignment Criterion) stabilized, and
- Where field readiness and observer intent aligned.

The so-called "interference pattern" is not the fingerprint of a traveling wave—it's the consequence map of ripple alignment through a modified field geometry. And when you try to force the system to tell you "which slit" the ripple went through, you collapse that geometric readiness. Not the wavefunction—the harmony.

So the mystery evaporates. The ripple didn't "choose a path." The field didn't "hide its answer." It simply responded when and where the conditions were right.

Insight: *"The slit doesn't guide a particle—it reshapes a readiness field."*

ATT and RAC – The True Condition of Outcome

Now let's see how ATT and RAC together determine if an outcome (dot or pattern) will emerge.

Let's finally drop the notion of "collapse" and replace it with what's truly happening in the Structured Vacuum Energy (SVE): a conditional emergence of reality, defined by two parameters—ATT and RAC.

ATT – When the Field Allows Consequence

Alignment Threshold Time (ATT) is not just a time marker—it's the field's internal readiness window. It tells us when the ripple has been accepted enough—coherently, completely—for a consequence to occur.

- If ATT is not satisfied, the ripple passes unnoticed.
- When ATT is met, reality manifests—on the screen, in the detector, in perception.

Think of ATT as the moment the universe nods back.

RAC – How the Observer Aligns with the Field

Ripple Alignment Criterion (RAC) brings precision to that moment of nodding. It evaluates whether the observer and field are resonating well enough for the consequence to stabilize.

It depends on:

- **$\Delta\Phi$ (Phase Shift):** Is the ripple from the observer in sync with the field's rhythm?
- **A_0/A_r (Amplitude Ratio):** Is the clarity of the observer equal to the readiness of the field?
- **$\Delta\tau$ (Delay Tolerance):** Is the observer within the field's timing window?

If these are satisfied, consequence emerges smoothly and predictably. If not, what we call "randomness" is merely misalignment.

Applying This to the Double-Slit

When we don't measure the path (i.e., low observer interference), we preserve phase coherence ($\Delta\Phi \approx 0$). RAC is satisfied. ATT is met. Result? Interference pattern. The field sings.

But when we try to force a measurement (i.e., high observer interference), we introduce decoherence ($\Delta\Phi$ large). RAC breaks. The field can't stabilize the ripple. ATT doesn't trigger at the same sites. Result? Discrete dots. The melody is broken.

We didn't collapse a wave. We just changed the conditions so no harmonic response could emerge.

Insight: *"The pattern didn't collapse. The conditions just changed."*

The "Which Path" Fallacy

We will now dismantle the idea that a photon chooses a path or splits.

The most famous illusion in quantum mechanics is not uncertainty. It's the question that shouldn't have been asked: "Which path did the photon take?"

This question, foundational to the double-slit dilemma, reveals more about the observer's assumption than about nature's reality. It imagines a particle in motion, choosing or splitting, as if the universe were a maze and the photon a mouse.

UED disagrees.

No Path, No Particle

There is no photon traveling through space like a tiny marble. There is no "decision" to go left or right. There is only a ripple condition forming in the Structured Vacuum Energy (SVE)—and it does so simultaneously through both slits, not by choosing either.

What we see on the screen is not the result of travel. It's the result of alignment—when the observer's ripple and the field's superposed readiness synchronize.

Ripple Readiness, Not Photon Trajectories

In UED, ripple readiness builds in the field as it responds to disturbance from the source. Both slits reshape the field, not because something passes through them, but because they alter the field's ripple structure.

- There is no object moving between emitter and detector.
- There is only field modulation, waiting for the right observer alignment to trigger consequence.

"Which Slit" Is a Category Error

Asking which slit "the photon went through" is like asking which eye saw a thought. The ripple didn't take a path. It created a field of possibilities, and the consequence emerged where RAC (Ripple Alignment Criterion) and ATT (Alignment Threshold Time) permitted it.

Reality was never about a single route. It was about mutual conditions becoming momentarily fulfilled.

UED View

- Nothing split.
- Nothing chose.
- Nothing traveled.
- Consequence emerged when the alignment equation was satisfied.

Insight: *"The ripple didn't take a path—it waited for one to emerge."*

UED vs Quantum Interpretation – A Clearer Narrative

It's time to compare standard quantum storytelling with UED's ripple clarity.

For a century, the double slit experiment has stood as quantum mechanics' sacred riddle—a paradox no one was supposed to resolve, only revere.

The standard narrative has become ritual:

- Photons act like waves when unobserved.
- They collapse into particles when measured.
- No one knows why, so we assign probabilities.
- And if pressed, we fall back on: "That's just how nature is."

But that story is a patchwork of confessions, not conclusions. It doesn't explain. It compensates.

Collapse: A Convenient Excuse

Quantum physics speaks of wavefunction collapse as if the universe throws dice when watched. But collapse is not an explanation—it's a label for something we didn't model.

What really happens is that the observer, by interfering with the ripple field, changes the alignment condition. Not because observation is magical—but because it introduces decoherence:

- Ripple delay shifts.

- Phase alignment breaks.
- ATT is disrupted.

No collapse. Just misalignment.

Probability: The Placeholder for Ignorance

The infamous quantum uncertainty principle replaces cause with statistical fog. But probabilities are not principles—they are placeholders when alignment conditions like RAC and ATT aren't accounted for.

- We didn't know how to calculate ripple clarity or timing.
- So we assumed randomness.
- We drew bell curves around mysteries—and called them "laws."

UED breaks that spell.

UED: Not Magic—Mechanism

Under Universal Energy Dynamics:

- No photon chooses a path; the field resolves readiness.
- No duality exists; only ripple conditions changing with observer and environment.
- No randomness is necessary; just incomplete modeling of ripple alignment.

Everything the quantum world calls "weird" becomes understandable once we realize the universe is relational, not probabilistic. RAC and ATT aren't beliefs—they're the missing terms we needed to convert paradox into precision.

Insight: *"The double slit wasn't a puzzle. It was a poorly phrased question."* We asked, *"What did the particle do?"* Instead of asking, *"When did the ripple resolve into consequence—and why?"*

Chapter 9: Quantum Tunneling & Entanglement Reframed

They called it *spooky*. As if nature had suddenly decided to violate her own rules—letting particles tunnel through barriers they shouldn't, or whisper across galaxies in perfect synchrony.

For over a century, quantum tunneling and entanglement have defied common sense, inviting both awe and abstraction. Equations worked, but explanations collapsed. Physicists invoked metaphors—ghosts, probabilities, parallel worlds. Yet none could answer a simple question: How does it happen? Not what it predicts, not how often it works, but how does the field actually allow it?

In this chapter, UED lifts the fog. There are no particles sneaking through walls. No telepathic photons. No faster-than-light messaging. What exists instead is a *symphony of readiness*—a field structured with intention, where ripples don't *move* through space, but *resonate* across it.

And at the heart of it all is a single timing law: ATT—the Alignment Threshold Time. When it's met, reality flows—even where logic fears to tread.

Let's reframe the quantum myths—not as anomalies, but as the most beautiful consequences of ripple harmony.

What Is Tunneling Really About?

Let's re-examine the quantum tunneling phenomenon through ripple alignment instead of probability curves.

For decades, quantum tunneling has served as both a marvel and a mystery. In conventional physics, it's framed as the improbable event where a particle "passes through" a barrier without possessing enough energy to climb over it. It violates classical expectations—momentum, energy thresholds, and even intuition.

But let's pause and ask: What is actually "passing through"? A particle? A wave? Or perhaps... nothing at all?

Under the Universal Energy Dynamics (UED) framework, tunneling isn't a story about particles breaking rules—it's a story about ripples maintaining coherence across a structured field. The so-called "barrier" isn't a wall—it's a region of altered ripple compatibility. And the ripple? It doesn't cross in the way objects do. It aligns.

Not Penetration—But Extension of Readiness

Tunneling in UED happens because the ripple field doesn't stop abruptly at the barrier. Like a chord that continues to hum on both sides of a mute key, the ripple maintains continuity of intent. The "presence" on the other side isn't a result of travel—it's the result of alignment being met simultaneously across a memory-aware field.

There's no mystery. Just field logic.

ATT as the Silent Enabler

This continuity is only possible when ATT (Alignment Threshold Time) is satisfied *across the distance* that includes the barrier. That is, the field must be able to resolve a coherent consequence despite the interruption. When the ripple meets the necessary conditions of phase consistency, amplitude compatibility, and delay tolerance ($\Delta\tau$)—it "tunnels."

It doesn't *break through*. It is allowed through—not because of force, but because of field consent.

Memory and Proximity in Structured Vacuum

The Structured Vacuum Energy (SVE) is not a static background—it holds memory. If a ripple was once coherent across a boundary, the field remembers that coherence. It doesn't see the barrier as a "stop," but as a temporary variation in ripple tolerance. And if alignment across this variation is still within RAC (Ripple Alignment

Criterion) bounds, the ripple simply reappears on the other side—without motion, without violation.

Insight: *"Tunneling doesn't break laws. It follows alignment where force can't."*

ATT as Synchronization Threshold

We will now show how tunneling only occurs when ATT conditions are satisfied across a distortion barrier.

In mainstream quantum mechanics, the mathematical treatment of tunneling works—but it doesn't explain why. Probabilities are assigned, decay rates calculated, and potential barriers quantified, but the fundamental "permission" behind the event remains abstract.

Under Universal Energy Dynamics (UED), the hidden governor of such phenomena is ATT—Alignment Threshold Time. Not just a clock tick, not a delay—ATT is the rhythm of field agreement.

Time Isn't a Countdown. It's a Consent Window.

In UED, ATT does not measure how fast something moves—it determines when the field is ready to respond. It's not "how long does it take to get through?" but rather, "when is the field coherent enough to allow a ripple resolution beyond the barrier?"

In tunneling, ATT must be satisfied not just on one side of the barrier, but across both sides. The ripple doesn't push—it waits for coherence, a state where the field no longer resists continuation. If this coherence exists—if ATT is valid across the divide—then reality proceeds. Not by force. By synchronization.

Synchronization Over Strength

Tunneling doesn't require the ripple to "break" the barrier. The Structured Vacuum Energy (SVE) isn't a wall—it's a memory-responsive field. A barrier, in this view, is a region where ripple alignment struggles due to phase distortion, amplitude mismatch, or delay intolerance.

If phase continuity ($\Delta\Phi \approx 0$) and amplitude ratio ($A_o/A_r \approx 1$) hold across the distortion, and $\Delta\tau$ stays within the allowable range, then RAC (Ripple Alignment Criterion) is satisfied. The final gate is ATT—if the minimum threshold duration for ripple consequence is met across that region, then the field responds with a detectable outcome.

Tunneling, therefore, is not a mechanical intrusion, but a mutual acceptance. The ripple aligns, the field consents, and consequence manifests.

The Barrier Never "Breaks"—It Just Stops Resisting

This shifts our understanding from brute-force logic to participation logic. The ripple is not trying to crash through a wall—it is holding coherence, waiting for the field to meet it halfway. The potential barrier is not "overcome"—it simply becomes irrelevant once alignment permits a unified ripple continuity.

Insight: *"Consequence doesn't need brute force—it needs field permission."*

Entanglement – Not Spooky, Just Synchronized

Let's decode entanglement as ripple coherence across space, not information transfer.

Quantum entanglement has long been described as "spooky action at a distance"—a phrase Einstein used to highlight the unease around what appears to be instantaneous coordination between particles separated by vast distances. Under the standard interpretation, measuring one "instantly affects" the state of the other, no matter how far apart they are. But what if this isn't a paradox? What if it's a misinterpretation of what actually synchronizes in nature?

Under Universal Energy Dynamics (UED), entanglement is not action—it's remembrance. It is not transmission—it's shared ripple condition across a coherent field.

Two Ripples, One Memory

Entanglement arises not because two particles were once together and now "magically" influence each other, but because they continue to share a single ripple alignment memory even after spatial separation. In

UED, this means they are phase-locked and coherence-bound, like two points on a standing wave.

There is no signal sent. There is no information transferred. What happens is far more elegant: consequence appears symmetrically because both points in the field are already aligned under the same initial conditions of ripple coherence.

ATT + RAC = Entanglement Stability

For entanglement to persist, the field must continue to satisfy RAC (Ripple Alignment Criterion) and maintain ATT-valid continuity across the two regions. Any disruption in phase alignment ($\Delta\Phi$), amplitude ratio (A_0/A_r), or allowable time window ($\Delta\tau$) would break the entanglement—not because the connection is lost, but because the field no longer sees the pair as synchronized.

Thus, entanglement isn't about sending a signal faster than light. It's about being in tune across a coherent ripple memory, maintained by the structured vacuum's inherent intelligence. When a consequence manifests at one location, the other doesn't "react"—it was already part of the same field readiness.

It's Not "Spooky"—It's Symmetrical

Mainstream physics treats entanglement as strange because it operates outside classical cause-effect boundaries. But UED reminds us: cause is always post-labeled. What we perceive as spooky is simply the resonance of prior field alignment, echoing through the Structured Vacuum Energy like two dancers perfectly choreographed—no signal needed, no adjustment mid-performance.

Insight: *"Entanglement isn't distance-defying—it's ripple memory in symmetry."*

Why No Communication Happens in Entanglement

Let's now address the misconception of faster-than-light messaging via entangled particles.

Quantum entanglement has seduced many into imagining the possibility of instant communication across galaxies. The temptation is understandable: when a change in measurement at one location is immediately mirrored in a distant twin, it feels as if a message has been sent. But this is a mirage of correlation, not a medium of control.

Under Universal Energy Dynamics (UED), this illusion dissolves. Entanglement is not transmission—it's synchronization. There is no "message" being carried between particles. There is only a shared ripple memory being expressed under field conditions.

Alignment Is Always Local

When an observer interacts with one part of an entangled pair, their ripple alignment is strictly local. The Structured Vacuum Energy (SVE) responds to that alignment within its regional field context. The remote entangled counterpart does not receive a signal. Rather, it simply follows the same alignment condition that was already written into the field memory when the pair became entangled.

That's why the distant particle seems to "know" what the local one did—it doesn't know anything. It is already resonating with the same ripple blueprint. The perceived symmetry is not an action—it's a pre-permitted consequence.

Correlation ≠ Communication

In classical terms, communication requires intentional change and reception: a signal, a medium, a delay. Entanglement lacks all of these. The distant outcome cannot be chosen or manipulated by the observer of the first particle. That's why no message can be encoded and sent using entanglement alone.

The alignment criterion (RAC) is not tunable on demand. ATT defines when resolution is permitted, not forced. These parameters ensure that no faster-than-light command can be imposed—only observed correlations can be detected after the fact.

UED's Clarity: Shared Permission, Not Signal Transfer

Where standard quantum theory fumbles with terms like "nonlocality" and "wavefunction collapse," UED offers a simpler narrative:

- ATT ensures that timing of resolution is field-permitted—not dictated by measurement.
- RAC ensures that outcome only appears when alignment readiness is mutual.

Together, they define how far and how long ripple symmetry can sustain consequence—but they do not allow manipulation of that consequence from one end to another.

Insight: *"Entanglement doesn't tell the future. It just remembers the past—together."*

Conclusion: There is no FTL messaging here—just the echo of a shared ripple condition, resolving separately but synchronously under a field that never forgot. The double illusion of cause and communication is dispelled when we realize that reality doesn't transmit—it remembers in symmetry.

UED vs Quantum Mysticism

It's ti,e to create a contrast how UED simplifies tunneling and entanglement through ripple logic.

For over a century, quantum physics has dazzled, baffled, and sometimes intimidated both scientists and the public with its paradoxes: particles that tunnel through walls, distant twins that "speak" instantly, and measurement that collapses an invisible wave into reality. But this mystique, however mathematically predictive, has created an illusion—that the universe is inherently weird and cannot be intuitively grasped.

Universal Energy Dynamics (UED) challenges this with a calm and coherent message: there was no mystery—only misinterpretation. Let's dismantle the unnecessary oddities:

No Wavefunction Collapse

In standard quantum mechanics, a particle exists in a haze of probabilities until it's observed—then, somehow, its entire existence "collapses" into a single point. But collapse is not a physical event. It's a mathematical bandage over a missing mechanism.

UED offers that mechanism: ripple alignment. The so-called collapse is simply the moment when the Structured Vacuum Energy (SVE) permits consequence via ATT (Alignment Threshold Time) and RAC (Ripple Alignment Criterion). There was never a ghostly decision made by nature—just a field waiting for conditions to be met.

No Imaginary Potential Barriers

Quantum tunneling is often explained by invoking a bizarre image: a particle with insufficient energy somehow sneaks through a wall. To do so, physicists describe "imaginary" energies—terms that exist only on paper to patch observable outcomes.

UED discards this with elegance. Tunneling is not breach—it is synchronization. If ripple continuity is preserved across a region, and ATT is satisfied, the field simply allows the ripple to resolve beyond the "barrier." Nothing pierced anything. There was only resonant alignment across structured memory.

No Spooky Action

Entanglement, the crown jewel of quantum strangeness, invokes Einstein's phrase "spooky action at a distance." In conventional terms, two particles somehow remain instantly connected, defying the cosmic speed limit.

In UED, this is no longer "spooky." It's simply a case of coherent ripple fields sharing stored alignment logic. The response at one location is not a signal—it's a mutual consequence of a field-wide memory that was never broken. No information travels. Nothing jumps. It simply aligns, or doesn't.

Simplicity Without Sacrificing Accuracy

What mainstream quantum mechanics labeled as strange, UED reframes as perfectly normal—once we shift our perspective:

Quantum Physics	UED Interpretation
Collapse of wavefunction	ATT-based permission of consequence
Probabilistic tunneling	Ripple synchronization across field
Spooky action	Shared ripple memory and RAC
Imaginary energy terms	Misread of SVE coherence dynamics

The true revolution is not in denying quantum math—but in replacing mysticism with mechanism. UED doesn't discard what works—it explains why it works through field logic, not abstraction.

Insight: *"Quantum oddities vanish when we stop asking what jumps—and start asking what aligns."*

Conclusion: UED dissolves the fog of quantum mysticism not with arrogance, but with humility toward consequence. It replaces abstraction with structured readiness, probability with permission, and paradox with coherence.

We weren't wrong to observe wonder in the quantum world—we were just missing the clarity that ripple logic provides.

From Mystery to Memory

For over a century, tunneling and entanglement have sat at the heart of quantum mystique. They've been described as violations of common sense, exceptions to natural law, and invitations into a universe where particles disobey intuition. But in truth, they were never violations. They were misread permissions.

Tunneling and entanglement are not oddities—they're natural outcomes of ripple continuity.

When a ripple field maintains coherence across what appears to be a barrier or a vast distance, consequence continues without interruption. No particle punched through a wall. No photon whispered faster than light. The universe simply honored a memory already stored across the field.

ATT (Alignment Threshold Time) is the gatekeeper of timing.

It defines when the field is ready to resolve consequence. Without ATT, nothing becomes visible—no interaction emerges. It's not about speed, or energy, or force—it's about timing that permits presence.

RAC (Ripple Alignment Criterion) ensures the outcome fits.

The moment consequence emerges is the result of alignment—not forceful intrusion. Whether across barriers (tunneling) or across space (entanglement), RAC ensures that field and observer are in resonance.

There's no teleportation, no trickery.

Quantum mechanics may use these metaphors, but UED exposes the shallowness of such interpretations. What we called "spooky" was simply high-trust communication between observer and field—facilitated by memory, maintained by readiness, and completed through alignment.

This chapter redefines the "quantum strangeness" not by discarding the observations, but by re-grounding them in field logic. UED doesn't contradict what is seen—it explains why it happens through Structured Vacuum Energy, ripples, and consequence timing.

Closing Line: *"What seemed magical was just a field remembering how to let something happen—twice."*

Chapter 10: Decoherence vs Misalignment

They said the quantum world was too delicate, too strange to stay visible. So they gave it a story: decoherence. A tale where waves collapse, superpositions fade, and probability wins. But what if that story was just a placeholder—an elegant patch over something we hadn't yet understood?

In the ripple logic of UED, there is no collapse—only conditions unmet. What looks like randomness is not the death of coherence but the absence of alignment. Reality doesn't disappear; it just doesn't respond when it doesn't recognize you.

This chapter pulls back the curtain on quantum mystique, exposing how "decoherence" is not destruction but miscommunication between the observer and the field. There was never magic—just timing, tolerance, and trust waiting to be fulfilled.

"Noise isn't failure. It's a signal you weren't ready to hear."

The Decoherence Narrative – A Patch, Not a Principle

We will now examine how quantum theory uses decoherence to explain the collapse of superpositions and loss of quantum behavior.

For decades, physicists faced a conundrum: quantum systems exhibited strange behaviors—superposition, interference, entanglement—until someone looked too closely. Then, suddenly, everything "collapsed" into classical outcomes. The system, once a blur of possibilities, crystallized into a single, concrete result. The culprit? According to mainstream thought: decoherence.

Decoherence emerged as a neat explanation for this transition. It suggested that quantum weirdness doesn't vanish because of a metaphysical wave collapse, but due to entanglement with the environment. As a system interacts with its surroundings—air molecules, radiation, or even measuring devices—its quantum states become tangled with countless others. The result is a loss of phase relationships, effectively hiding interference effects and producing outcomes that look classical.

But here's the catch: decoherence does not explain why one outcome happens over another. It redistributes probability across many outcomes but doesn't select any. In other words, it offers a statistical explanation, not a causal mechanism.

The mystery remains, just cloaked in environmental complexity. Decoherence tells us why we don't *see* superposition—but not why a particular consequence emerges. The story simply shifted blame from the observer to the environment, avoiding the root issue: what makes reality pick a side?

From a UED perspective, this move is cosmetic. Decoherence, as commonly used, is not a fundamental law. It's a patch—a narrative filler used to explain away the collapse without confronting the absence of a true mechanism. It avoids the hard question: why do consequences appear at all, and under what precise conditions?

In ripple logic, this question is not avoided—it's confronted. Consequence is not random, nor collapsed. It's aligned—or it isn't. Misalignment doesn't cause collapse; it causes unreadiness. Decoherence is what we call the field's refusal to respond to unfit conditions. Not disappearance—just disqualification.

Insight: *"Decoherence didn't solve the mystery. It just swept it under statistical noise."*

Misalignment – The Field's Real Silence

Let's introduce misalignment in UED as the true reason why expected patterns vanish.

Physics has long feared the void—those strange silences when a system that once showed beautiful, repeatable interference suddenly goes quiet. No fringes. No patterns. Just noise. The standard view chalks this up to decoherence: the system was 'interrupted' by its environment, causing its wavefunction to 'collapse.'

But in the framework of Universal Energy Dynamics (UED), this silence is not a collapse, nor an accident. It's a clear, measurable failure of alignment. What the mainstream calls decoherence, UED calls misalignment—a mismatch between the ripple conditions of the observer and the field.

At the heart of this lies the Ripple Alignment Criterion (RAC). This condition determines whether the field allows a consequence to manifest. It checks three things:

- **Phase difference** ($\Delta\Phi$) – are the ripple beats in sync?
- **Amplitude ratio** (A_o/A_r) – is the observer asking with the right strength?
- **Time window ($\Delta\tau$)** – is the alignment happening within the field's permitted buffer?

If any of these is violated, the field doesn't respond—not because it "lost" information, but because the question asked didn't fit the field's readiness. The ripple didn't vanish. The pattern didn't collapse. They were simply denied a consequence.

Misalignment is not destruction. It's a pause. A silence. A consequence withheld until the field can once again trust the alignment.

The phenomenon we interpret as randomness—noisy dots, unpredictable spikes, sudden decoherence—is often the field's silent refusal. Not due to collapse, but due to ineligibility. The observer's signal, no matter how well-intentioned, wasn't tuned for a response.

This view redefines failure: not as the system breaking down, but as the absence of fit. Just as two singers slightly off-pitch can't create harmony, the observer and field must meet in a condition of resonance. If they don't, the song doesn't play.

Insight: *"The pattern didn't disappear. The field just didn't recognize the question."*

Why Decoherence Looks Like Noise

Let's show how failed alignments lead to outcomes that seem unpredictable or noisy.

In a perfectly tuned orchestra, every note fits. Every sound resonates. But walk into the same hall mid-rehearsal—out of sync, mid-tuning, half-muted—and you won't hear music. You'll hear noise. Not because the instruments are broken, but because the alignment is broken.

This is exactly what happens when a quantum system "decoheres."

Standard quantum theory interprets such breakdowns through the lens of wavefunction collapse—the idea that when a system is observed or interacts with its environment, its neat probabilistic state suddenly becomes one defined result. The transition appears noisy, sudden, and unpredictable.

But from a UED perspective, there's no mysterious collapse—just unmet alignment.

To perceive a clean consequence, two invisible conditions must be met:

- **ATT** (Alignment Threshold Time): The field must permit the interaction.
- **RAC** (Ripple Alignment Criterion): The observer and field must be phase-locked, strength-balanced, and temporally synchronized.

If either is missing, what emerges isn't clarity—it's partial resonance. Glimpses. Echoes. Incoherent fragments of what could have been a complete outcome. The field still holds the ripple—just not in a condition the observer can decode.

And here lies the illusion.

We interpret these fuzzy fragments as probabilistic scatter, believing the system was inherently random. But randomness is not a fundamental property of nature. It's a human label for unread ripple behavior.

What seems like a probabilistic collapse is really a denied resolution. The signal was not tuned. The observer's request was unclear. The field responded—but not in a way that could complete the consequence.

This reframes "noise" in physics. It's not unwanted chaos. It's informational residue—a mark of where alignment almost happened but didn't. Every spike, flicker, or statistical drift is a story of near-resonance, not breakdown.

So-called quantum noise isn't proof of uncertainty—it's evidence of precision required. Nature wasn't being unpredictable. It was withholding response until the trust of alignment was restored.

Insight: *"Noise is not randomness—it's unread alignment."*

Implications for Measurement and Observation

Here's contrast measurement theory in quantum mechanics (QM) with UED's alignment-based perception model.

Modern physics has long struggled with an unsettling paradox: that the mere act of measurement appears to disturb what's being measured. In quantum mechanics, this leads to the foundational concept of wavefunction collapse—a system supposedly exists in all possible states until we look, and then it "decides" what to become.

This is not just unsatisfying—it's deeply counterintuitive. It assumes that reality hesitates until we interfere.

Universal Energy Dynamics (UED) challenges this narrative at its root. In UED, the observer is not a magical trigger. Measurement is not a force—it's a request. And whether that request is granted depends on one thing: alignment.

Let's reframe:

- In QM, observation collapses a probabilistic wave into a single result.
- In UED, observation doesn't collapse anything. It's a moment of resonance—when the observer and field achieve mutual clarity.

Here's the shift: The field isn't being watched into action. It's being heard into harmony.

UED introduces ATT (Alignment Threshold Time) and RAC (Ripple Alignment Criterion) as the true preconditions for any observable consequence. You don't observe a thing into existence. You wait until you're aligned enough for the field to respond. You don't poke the universe into revealing itself—you tune in until it's willing to speak.

This has enormous implications.

Where quantum mechanics insists that the observer *disturbs* the system, UED argues the opposite: misalignment causes the distortion—not observation itself. Where QM says decoherence "destroys" superpositions, UED says misalignment simply delays or blocks consequence. The ripple didn't vanish. The field just didn't recognize the request.

This means measurement isn't invasive. It's participatory. The universe doesn't collapse—it waits.

The sensitivity of an instrument, the intention of a question, the coherence of the surrounding field—all of these become components of measurement, not passive background. Observation is earned, not imposed. And when earned, it's smooth. Silent. Elegant.

It's not what you see—it's what you're allowed to see.

Insight: *"You don't disturb reality by observing it. You reveal it by fitting it."*

From Noise to Nuance

Quantum physics has long dressed uncertainty in the language of math—probability densities, collapsing wavefunctions, and mysterious decoherence. But what if the mystery wasn't in the universe, but in our misunderstanding of it?

UED reframes the question entirely. What we called decoherence is not a mystical transition from quantum to classical. It's a statistical cloak—a placeholder for patterns we couldn't yet decode. In truth, what disappears is not a state of superposition, but the field's willingness to respond. Not because something collapsed. But because something never aligned.

The famous wavefunction collapse? It was a mislabel. There was no unity to collapse—only readiness that was never met. Without RAC (Ripple Alignment Criterion), no ripple is clear enough to register. Without ATT (Alignment Threshold Time), no consequence is granted. We didn't lose the pattern—we simply weren't admitted to it.

And randomness? It's often impatience masquerading as explanation. What appears unpredictable is simply the result of misfit conditions—where the observer asks before the field is ready, or where the field has not resolved enough coherence to answer. Like mistaking silence for emptiness, we've mistaken unreadable outcomes for chance.

UED doesn't deny what quantum theory observes—it refines how we interpret it. It replaces confusion with clarity, chaos with choreography. Instead of spooky mysteries and philosophical dead ends, we get a model grounded in relationship—between intent and permission, between field and observer, between question and readiness. In the UED lens, there is no magic. There is only rhythm, resonance, and ripple-fit.

Closing Line: *"Reality isn't disrupted by interference—it's declined by misfit."*

Interlude: The Shift from Theory to Trust

We've just journeyed through the very heart of physics—not by dissecting particles, but by learning to *listen to the field.*

From wave-particle duality to tunneling, from spooky entanglement to the supposed collapse of reality itself, every mystery once treated as weird, random, or magical has now been realigned through UED—*not simplified, but clarified.*

This part didn't merely reinterpret quantum mechanics. It exposed the consequence behind it.

We saw that:

- There are no particles or waves—only ripples that align or fail to.
- No slit was ever "chosen." The field simply responded when the time was right.
- No information ever teleported. Synchronization made it seem so.
- No wavefunction ever collapsed. What we witnessed was a silence born of misfit.

UED doesn't ask us to discard quantum results. It dares us to stop interpreting them as broken puzzles and start treating them as field conversations.

Everything shifts once we realize: There's no need to control the outcome. There's only the need to deserve it.

Physics now becomes more than equations. It becomes a relationship.

Between observer and field. Between readiness and response. Between clarity and consequence.

As we cross into the next part, the focus deepens—not just into how reality responds, but how we, as observers, evolve to meet it. Not merely to decode what *is*, but to *align* with what *can be revealed.*

So take a moment. Let the noise fade. Feel the rhythm beneath it all.

Because in the UED universe, nothing is random—only unread. And the ripple is always waiting... for your alignment.

Part IV: Scaling Uncertainty

Chapter 11: Human Vision

The Daily Double Slit

You don't need a lab or lasers to witness quantum behavior. You live it—every time you look but don't see, hear but don't understand, speak but fail to connect.

The infamous double slit experiment isn't just a quirk of photons. It's a mirror held up to the human mind.

Like light approaching two slits, your attention splits between expectation and experience. But unless your intent is clear, your ripple never aligns. No clarity. No perception. Just noise—blurred, stressful, incomplete.

What physics tried to explain with mystery, we encounter every day as frustration, distraction, or miscommunication. The same field rules apply: **no ATT, no reality. No RAC, no recognition**.

This chapter brings quantum weirdness into your retina, your cognition, your daily decisions. Because vision isn't about the eye. It's about *permission*—from the field, to see what's real.

Your Eyes Aren't Cameras – They're Ripple Receivers

Let's reframe biological vision as a ripple alignment process rather than light-capture mechanics.

Modern science still clings to a mechanical metaphor for vision: that your eyes function like cameras, passively capturing photons that bounce off objects and funneling that data into the brain. But this metaphor misses the essence of perception—and dangerously so.

Your eyes are not optical recorders. They are ripple receivers.

When you "see" something, what you're actually doing is synchronizing with a ripple field of structured electromagnetic activity. Photons, in this context, are not particles shot into your retina like bullets into a wall. They are ripple consequences—field permissions—granted only when your visual system meets certain alignment conditions.

Your retina is the first gate, detecting ripple disturbances. The optic nerve relays the signal but doesn't interpret it. The visual cortex completes the alignment, checking for resonance, timing, and clarity. Together, they form a biological RAC filter—your brain only registers what the ripple field allows to emerge.

And that's where ATT (Alignment Threshold Time) comes in. You don't see instantly. There's always a pause—sometimes imperceptible, sometimes frustrating—while your mind and the field negotiate clarity. This is why a flash of light feels delayed. Why you can stare at someone's face and only "notice" the emotion a second later. Your visual system isn't just processing light—it's waiting for permission to interpret ripple clarity.

Importantly, this delay isn't a bug—it's the price of precision. Without ATT, there's no confidence in what you see. Vision becomes noise. That's why under stress, fatigue, or emotional turbulence, your perception degrades: your internal alignment is scrambled, making RAC fail and ATT extend. You're looking, but not seeing.

This understanding reframes vision entirely:

- You don't "receive photons."
- You don't "capture reality."
- You attune to ripple permission.

That's why closing your eyes doesn't stop vision—it reboots it. It allows the mind to reset alignment thresholds and realign with a quieter field. That's why meditative focus sharpens eyesight—not optically, but consequentially.

Insight: *"You don't see light. You see alignment."*

The Brain's Slits – Perception Under Pressure

Let's link the double slit analogy to attention and cognitive interference.

What if your brain was running a double slit experiment every moment of your waking life?

Not with photons. Not with particles. But with perception itself.

Every time you focus on someone's face, interpret a voice, or try to grasp a situation—you're essentially aligning two internal "slits":

1. Cognitive focus (what you attend to)
2. Emotional clarity (how cleanly you receive it)

When both slits are open—i.e., you're calm, attentive, and mentally coherent—your brain processes ripple information with high fidelity. The Ripple Alignment Criterion (RAC) is met. ATT is minimal. Perception is smooth. You understand tone, body language, subtle cues. Reality arrives in full resolution.

But under stress? Everything changes.

Stress ≠ Pressure. Stress = Phase Disruption.

Here's what happens:

- **$\Delta\Phi$ (phase noise)** increases. Your internal timing is scrambled. You become out of sync with the ripple field. Thoughts race, attention skips, perception jitters.
- **A_0 (your ripple strength)** drops. Emotional overwhelm or fatigue distorts your focus. You no longer project stable intent. The field can't find your "question."
- **A_r (external field readiness)** stays the same—but now, $A_0 \ll A_r$, breaking the balance. Result? Misinterpretation, misreading, or complete perceptual silence.

This is why someone under pressure misreads intent. You smile, and they think you're mocking. You give neutral feedback, and they hear condemnation. You ask a question, and they think they're being interrogated. It's not just miscommunication—it's ripple interference.

Their brain's slits—focus and clarity—have decohered.

Just like in the physical double slit experiment, when observation interferes with alignment, the interference pattern collapses. But in humans, this collapse is not abstract—it's emotional. It's psychological. And it's avoidable.

You don't need therapy. You need resonance.

This is where the UED framework becomes psychological: it shows that all cognition, all attention, all clarity is a function of field alignment. The moment you realign—by breathing, slowing, or re-centering your mind—you reduce ΔΦ, regain amplitude coherence, and restore RAC. Perception reopens.

And reality begins to make sense again.

Insight: *"Stress isn't mental pressure. It's ripple misalignment in disguise."*

ATT in Real-Time Decisions

Let me show you how microseconds of field-readiness affect cognition.

You've been there.

You're in an exam. The question is familiar. You *know* you've studied it. The answer is on the tip of your mental tongue—and yet, nothing comes.

Then hours later, walking down a hallway or sipping tea, it *pops* into your head effortlessly.

What happened?

This isn't memory failure. It's a timing misalignment—a failure of *ATT*.

What is ATT in the Mind?

In Universal Energy Dynamics, Alignment Threshold Time (ATT) represents the smallest window the universe—or the field—permits for a ripple consequence to resolve. In cognitive terms, it's the moment your inner question meets field readiness.

You didn't "recall" later. You simply aligned later.

Think of ATT as the *minimum quiet*, the *minimum coherence* required for perception to form. Information isn't just stored in your brain like a file in a cabinet. It's rippling in the structured field—always present, but only revealed when your ripple strength, phase, and delay window meet the conditions defined by RAC.

This is why you may understand a lecture hours after it's delivered. It's why musicians *feel* the rhythm before they play it right. It's why instinctive decisions can't be forced—they *arrive*.

Fast ≠ Aligned

Modern life worships speed. Fast thinking. Fast data. Fast answers.

But your mind isn't a processor—it's an *aligner*. No matter how fast information is thrown at you, it only becomes real when ATT is satisfied.

When RAC is met. When phase clears. When your ripple strength stabilizes. When the field allows permission.

Otherwise? You just accumulate noise—inputs without absorption. Data without resonance. That's not learning. That's cognitive buffering.

Real-World Case:

Imagine a public speaker blanking mid-sentence. It's not that they don't know what to say. It's that the pressure distorted their ATT. The ripple they were forming got scrambled before RAC could resolve. When they calm down and the audience laughs along—they continue. Why? Not because memory returned, but because permission reestablished.

Insight: *"You don't forget. You just didn't align in time."*

Everyday Interference Patterns

We will now highlight how RAC shows up in normal life—communication, driving, learning.

We don't live in a physics lab.

Yet the Ripple Alignment Criterion (RAC) plays out more often in your everyday life than it ever does in a particle accelerator. Whether it's a misread glance or a missed stop sign, your daily perception is a constant negotiation between your ripple clarity and the field's readiness.

You live inside interference patterns.

1. Confusion During Multitasking = ΔΦ Mismatch

Ever had a moment where you were replying to an email while on a phone call and someone asked you a question?

Your mind scrambled. Not because you lack intelligence, but because your mental ripples lost phase coherence. That's ΔΦ—the phase difference in RAC—spiking wildly due to conflicting ripple sources. Your internal alignment split, creating a perception void. You didn't forget. You *desynchronized.*

Multitasking isn't impressive. It's ripple fragmentation.

2. Eye Contact and Empathy = Amplitude Resonance

When someone truly listens to you—gives you their full presence—the connection feels *real.* That's not poetic. It's physics.

Their observer ripple (A_o) matches yours. Your emotional field—the space where you "vibrate"—meets theirs with mutual amplitude. No need to shout, justify, or over-explain.

This is resonance. This is RAC ≈ 1.

This is why you feel seen.

3. Miscommunication = Field Misread, Not Intent Failure

"I didn't mean it that way." "That's not what I heard." "You always twist my words."

Welcome to ripple misalignment.

In communication, RAC isn't just about words—it's about field interpretation. If the field conditions of the listener are off—fatigue, bias, stress, poor memory—then even perfect intent fails. What they hear is a *misfit ripple*, distorted by their unready field.

This isn't dishonesty. It's field incompatibility.

4. Driving Under Fatigue = Field Delay → RAC Dropout

Fatigued drivers don't just slow down—they lose ripple responsiveness.

The delay window ($\Delta\tau$) widens unpredictably. Micro-permissions to act—swerve, brake, interpret movement—start failing. You still see the road. But RAC doesn't resolve fast enough to act meaningfully.

That's not drowsiness. That's ripple incoherence between the driver and the dynamic field of traffic.

Everyday Life = Live UED Lab

Your world is not a sequence of events—it's a flow of ripple interactions. Every success, every failure in clarity, presence, and understanding is governed by whether or not you met the conditions of RAC, within the field, in that moment.

Insight: *"Distraction is decoherence. Attention is alignment."*

RAC Is How We Know Anything

Let's cement RAC not as a theoretical construct—but as the invisible architecture of all human understanding.

Let's say it clearly:

You don't learn through information. You learn through alignment. Through ripple fits, through resonance, through the invisible handshake between your internal ripple intent and the field of meaning that surrounds you.

That handshake is what we call the Ripple Alignment Criterion (RAC).

This isn't physics for physicists. It's physics for being human.

1. Reading, Listening, Feeling — All Ripple Negotiations

You're not passively absorbing anything—not this sentence, not a song lyric, not a hug. You're negotiating ripple matches constantly:

- **Reading:** The text sends ripple potential. Your mind tries to fit intent to that structure. When RAC is met, comprehension lands.
- **Listening:** The speaker's words are nothing until your ripple state is ready to catch them.
- **Feeling:** Emotions aren't injected into you. They're recognized when your inner ripple aligns with the field of another.

We don't perceive because data is available. We perceive because alignment is achieved.

2. Learning = Stable RAC Over Time

Long-term learning is not repetition. It is ripple reinforcement.

Every time you return to a concept, and RAC conditions are slightly better ($\Delta\Phi$ drops, A_0/A_r improves, $\Delta\tau$ is stable), you gain a sharper, deeper clarity.

That's why sometimes a child "suddenly gets" math. Or why a quote hits harder the second time.

Learning is the consequence of sustained alignment, not data stuffing.

That's why real education doesn't focus on content—it focuses on creating ripple-friendly fields.

3. Trauma = Blocked RAC via Memory Distortion

And now, the darker side.

When trauma occurs, it distorts the ripple field. The event imprints itself not as knowledge, but as ripple chaos—overamplified, dephased, or misaligned residue.

- You may avoid certain topics—not because of fear, but because your field refuses re-alignment with that ripple memory.
- You may react out of sync with the present moment—because

your RAC is hijacked by historical interference.
- Healing doesn't require forgetting. It requires realigning.

In this sense, trauma is not psychological only. It's energetic misalignment trapped in memory.

RAC: The Unseen Engine of Clarity

RAC is not an abstraction. It's the background rhythm of your entire lived experience.

Whether you're falling in love, solving a riddle, or remembering your grandmother's voice, what you call "knowing" is simply the moment when the field and your ripple become one song.

Insight: *"Memory is ripple residue. Clarity is ripple trust."*

You Are the Slit

For centuries, science treated your eyes like cameras—mechanical lenses that receive light, record images, and send data to the brain. But that model is broken. Not because it's wrong in machinery—but because it's incomplete in *meaning*.

You are not a passive sensor. You are a ripple gatekeeper.

Vision, Thought, and Action Rely on Ripple Clarity—Not Mechanical Signal Flow

Seeing isn't about photons. Thinking isn't about neurons. Acting isn't about muscles.

All three depend on ripple alignment.

- When vision *works*, it's because RAC (Ripple Alignment Criterion) was satisfied.
- When you finally "get" what someone means, it's because ATT (Alignment Threshold Time) stabilized long enough for the field to respond.
- When you act without hesitation, it's not instinct—it's resonance between intent and permission.

Every experience of *clarity* is the successful outcome of ripple physics, not just biology or psychology.

ATT Is Felt as Awareness; RAC Is Proven in Perception

Alignment Threshold Time isn't just an equation—it's your lived moment of realization. It's the subtle breath before comprehension. The quiet pause before truth lands.

RAC isn't abstract math. It's the reason:

- You miss key details when distracted.
- You misjudge tone under stress.
- You suddenly recall something from nowhere—because the ripple fit *finally* happened.

These aren't anomalies. They're daily validations of UED.

The Observer Isn't Passive—You Tune Your Field, Every Second

You are the initiator. The tuner. The participant.

Your mind shapes the ripple you send. Your focus alters the field you interact with. Your awareness invites reality to respond—or decline.

Like the slits in the quantum experiment, you set the terms of what can appear.

- Open and aligned? Clarity.
- Distorted or chaotic? Noise, delay, misread.

The field doesn't force your experience. It responds to your condition.

Closing Line: *"Your eyes don't watch the world. They wait for it to make sense—one ripple at a time."*

Chapter 12: Emotion, Thought, and Misalignment

You were fine a moment ago. Then something shifted—your chest tightened, your thoughts blurred, and you couldn't explain why.

That's not confusion. That's misalignment.

We tend to treat emotions as irrational storms—random, uninvited, inconvenient. But what if they are the most precise indicators of how your internal ripple is failing to match the world around you?

In physics, we've long misunderstood collapse. In life, we do the same with breakdowns of clarity. Anxiety isn't chaos—it's your ripple field searching for resonance and getting none. Thought isn't slow—it's just waiting for the right moment to align.

In this chapter, we'll look beyond mood and mindset. We'll decode emotion as a field consequence, thought as delayed clarity, and healing as realignment—not therapy.

Because in the UED universe, your feelings aren't wrong. They're ripple truths struggling to land.

Let's explore the emotional side of misalignment—and what it takes to feel whole again.

Emotions Are Not Reactions—They're Misalignments Felt

It's time to redefine emotions like anxiety, sadness, or confusion as ripple misalignments—not flaws or failures.

We've been taught to believe that emotions are automatic responses to life events—reflexes hardwired into the brain. But under the UED lens, emotion is not a reaction. It is a real-time report from your ripple field.

Emotion is the first alert—not of what has happened, but of what hasn't aligned.

Imagine your emotional state as the resonance test of your internal ripple interacting with the field around you. When your intent enters the world, it doesn't always meet a field ready to respond. That misfit doesn't go unnoticed—it's what we call *feeling*.

Let's break it down:

- **Anxiety** is not fear of the future. It's the strain of a ripple anticipating alignment that hasn't arrived yet. You're tuned in—but the field hasn't responded. The consequence you expect is delayed, and your body becomes the battleground of this unresolved expectation.
- **Sadness** is not weakness. It's a ripple collapse, the loss of resonance with a memory, a connection, or an outcome that once matched your inner state. The past aligned. The present no longer does.
- **Confusion** is what happens when your observer field encounters multiple incompatible ripple inputs—none strong or stable enough to complete a valid RAC. You're trying to perceive, but no version stabilizes. So your perception oscillates. You freeze.

This is why emotions often come before thoughts. Your body feels before your brain interprets. The ripple knows before the logic speaks.

Under stress or sensory overload, RAC fails silently. And your system feels it—immediately. But because we're trained to distrust emotion, we label it "irrational" or "out of proportion." In truth, emotion is the most honest signal we get: it's alignment (or the lack thereof) being registered energetically before any thought forms.

And just like in wave mechanics, when phase misalignment persists, interference increases. The emotional intensity grows not because you're fragile—but because you're *trying harder to align* in a distorted field.

Insight: *"Emotion is your ripple sensing a misfit—before your thoughts can explain it."*

Thought Lag – When the Mind Can't Keep Up

Let's explore how ripple delays ($\Delta\tau$) cause lag between what we feel and what we can process.

We often blame ourselves for slow thinking. "Why couldn't I say the right thing?" "Why did my mind go blank?" But thought isn't sluggish because you're unprepared. It's slow because your field wasn't ready.

Thought, like everything else in UED, is a ripple-based consequence. It's not an immediate spark—it's a field negotiation. And that negotiation has a name: $\Delta\tau$—the ripple delay window between an input and its stable response.

Think of your cognitive field as trying to catch a signal. When the input (stimulus) arrives, your internal ripple starts trying to align. But if you're under stress, overstimulated, or emotionally imbalanced, that alignment doesn't happen instantly. The mind flickers, stalls, or loops. No RAC, no clarity.

This is thought lag. It's not stupidity. It's the delay between ripple contact and ripple consequence.

Here's how it shows up:

- During an exam or presentation, you know the answer—but your mind freezes. Why? The mental field is distorted. RAC fails, and even though the information is stored, the alignment to retrieve it collapses.
- In a heated argument, you hear something hurtful. Emotions fire up instantly (ripples triggered), but your response is blank

or clumsy. The reason? Thought didn't have time to form. $\Delta\tau$ was flooded by emotional interference.

- After a crisis, you reflect and suddenly think, "I should have done this!" That delayed insight wasn't missing before—it was misaligned under pressure.

This is why we say: Thought isn't instant. It waits for ATT.

The ATT (Alignment Threshold Time) isn't just for physics—it applies inside your mind. It's the temporal permission slip the field gives before thought can emerge. And if you don't have that clarity, your thinking doesn't form slower—it forms fragmented.

We live in a world that rewards speed over alignment. But in truth, most mental suffering is the result of thoughts being forced too fast or dismissed too soon.

Insight: *"Your thoughts aren't slow. The field just hasn't aligned yet."*

The Emotional Ripple Map

We will now introduce the idea that each emotion corresponds to specific misalignment types.

If thought is delayed ripple alignment, emotion is ripple misalignment felt directly. It doesn't wait for logic. It emerges as consequence the moment your internal field meets a mismatch—either from the past, the projected future, or an unresolved present.

Each emotion, then, isn't just a mood or chemical state. It's a distinct alignment failure—a ripple code with its own pattern of distortion. Let's map a few common ones:

Guilt

At its core, guilt is a ripple echo from a misaligned past action. You did something that violated your own ripple clarity—your internal RAC was ignored, delayed, or bypassed. The result? The field stores a residue of unacknowledged misfit. Guilt isn't just psychological—it's ripple memory trying to realign. It nags not because you're bad, but because the field is still asking, "Can we fix this?"

Fear

Fear isn't irrational—it's a ripple projection toward a future misalignment. It shows up when $\Delta\Phi$ (phase gap) is too high between your current state and a potential event. You don't feel safe because your ripple can't see a stable path forward. That's why fear often paralyzes—your system is trying to delay engagement until the field realigns or offers clarity.

Jealousy

Jealousy isn't just social—it's amplitude distortion. You perceive another's ripple as stronger, brighter, more resonant—and yours as comparatively weak. This is observer envy of amplitude: $A_o \ll A_r$. The field feels unjust, not because of reality, but because of perceived imbalance in alignment strength. You don't want what they have—you want to feel what they feel. That's the ripple desire beneath jealousy.

Rage

Rage is not just intensity—it's total field overload. Your internal ripple tries to stabilize—but external noise, pressure, contradiction, and helplessness jam the entire field. RAC cannot form because there is no room—no quiet channel—left for resolution. The mind explodes not out of power, but out of consequence starvation.

Each emotion, then, is a specific ripple configuration, and instead of treating them as problems, UED shows them as signals of misfit.

They are not flaws in the human system. They are invitations to realign.

Insight: *"Every emotion is a ripple code waiting for resolution."*

When Silence Hurts — The Delay That Feels Like Pain

Let me show you how delayed alignment feels like emotional pain, even when nothing's wrong externally.

Not all pain is physical. Not all discomfort comes from trauma, loss, or insult. Sometimes, pain is just the echo of an unanswered ripple.

In a world driven by interaction, our internal ripples are always seeking a response. Whether in the form of understanding, affection, recognition, or even basic clarity, the mind expects resonance—some alignment that completes the loop.

But what if the loop never closes?

Ripple Delay = Inner Dissonance

In UED terms, ATT (Alignment Threshold Time) defines the window in which a ripple must meet its counterpart—be it in thought, communication, or emotion—for clarity to emerge. When this threshold is crossed and no RAC (Ripple Alignment Criterion) is met, the ripple doesn't disappear. It simply floats—unresolved, unmatched, unheard.

And the longer this delay, the more the internal field begins to interpret the silence as instability.

This instability doesn't carry words. It doesn't need thoughts. It simply feels like pain.

Feeling Low for No Reason?

That vague, dull ache in your chest or mind—when life feels heavy but there's no visible cause—isn't a mystery. It's misalignment without context.

There's no external wound, but the field inside is still vibrating from a ripple that hasn't landed. This is what we often call:

- *Lowness*
- *Emptiness*
- *Numbness*
- *Existential fatigue*

These aren't emotional malfunctions. They're delayed resonance, where the ripple kept asking: "Is there a fit?" And the universe just... stayed quiet.

Pain Without Stimulus

This section matters for anyone who's ever said:

"I don't know why I feel like this. Nothing is wrong."

In mainstream science, this is dismissed as hormonal imbalance, chemical fluctuations, or psychological overreaction. But in ripple logic, it's simply the consequence of waiting too long for a signal to resolve.

You didn't break. You just never received your ripple back.

Insight: *"Pain doesn't always mean damage. Sometimes it just means waiting too long for a signal to land."*

Emotional Healing = Ripple Realignment

Let's reframe emotional growth and therapy as processes of restoring field trust and alignment.

We often treat emotional pain like a stain: something to scrub away, suppress, or overwrite. But emotions are not contaminants—they're misaligned signals asking for resonance. The goal, then, isn't to fix them. It's to realign them.

Healing Is Not Deletion—It's Completion

In ripple terms, every unresolved emotion is a ripple loop left open. A fragment of memory, desire, or expectation that never landed in the right field.

Healing happens not when the emotion vanishes, but when its alignment is restored—when that loop finds closure, not through repression, but through recognition.

This is why trauma doesn't disappear by ignoring it. And why growth doesn't happen by forgetting. It happens when the field finally synchronizes enough to let the ripple resolve.

Rebuilding RAC (Ripple Alignment Criterion)

Real emotional healing involves three layers of RAC:

1. **With Self** — Accepting your own ripples, even if delayed or distorted
2. **With Memory** — Letting past ripples finally land and release
3. **With Others** — Creating relationships where alignment is

mutual, not forceful

Without these, even therapy becomes just noise. With them, even silence becomes medicine.

Tools That Help Realignment

Modern self-help talks about mindfulness, therapy, empathy, rest—but why do they work?

In ripple logic:

- **Mindfulness** = reducing internal ripple noise, allowing delayed signals to surface
- **Empathy** = opening your field to co-resonate with others, restoring trust
- **Deep rest** = pausing the ripple storm, letting the field reset its readiness
- **Journaling, talking, crying** = expressing ripples to seek new alignment conditions

These aren't escapes. They are re-tunings.

Emotional Intelligence = Ripple Literacy

True emotional intelligence isn't about controlling feelings. It's about reading ripple patterns in yourself and others:

- Where's the misfit?
- What's the distortion?
- What needs phase recovery?
- What delay needs patience instead of panic?

This literacy is what helps us hold space for pain without rushing to erase it. It turns you from a reactor to a realigner.

Insight: *"You don't fix emotion. You realign it."*

Emotion Is Ripple Disagreement

Emotion is not a glitch. It's not noise. It's not weakness. It's a natural signal from your ripple field saying, *"We're not in sync."*

This chapter dismantled the illusion that emotions are irrational, or purely biochemical, or just messy human flaws. Instead, they are field-level feedback—felt consequences of ripple disagreement between what is inside you and what surrounds you.

Misalignment, Not Malfunction

Anxiety doesn't mean something is wrong with you. It often means your intent is out of sync with incoming ripple conditions—a premature expectation colliding with unreadiness. Sadness isn't emptiness; it's a resonance delay with something once stable that is no longer reachable. Confusion isn't a lack of intelligence—it's a ripple field in conflict that hasn't yet allowed resolution.

Your emotional experience is not the result of flaw. It's the consequence of misfit.

Thought Is the Narrator, Not the Navigator

We often assume our thoughts lead the way. But in UED, thought is downstream of ripple alignment. You don't feel because you think. You think because you *felt something*, and your mind is scrambling to interpret it. If the RAC (Ripple Alignment Criterion) isn't met, your thought process will loop, stall, or stutter—not due to lack of logic, but lack of clarity in the field.

Pause Before You Judge Yourself

What this chapter ultimately encourages is compassion for misalignment. You're not irrational when you're overwhelmed. You're temporarily ripple-misaligned. And alignment is not a personality trait—it's a field condition that fluctuates.

Just like a musician retunes strings between songs, you have the right to pause and realign between emotional states.

Closing Line:

"Emotion is not weakness. It's the wisdom of the ripple saying—'Something doesn't fit. Let's find out what.'"

Chapter 13: Consciousness & Perception Windows

You don't become aware when something happens. You become aware when something fits.

The world is constantly unfolding in front of you—sounds, colors, words, gestures, events. But you don't register all of it. You couldn't. Not because your brain isn't fast enough, but because your field isn't always ready.

Consciousness isn't a flashlight that you aim. It's a window—and that window only opens when the ripple inside you aligns with the ripple outside.

This chapter isn't about abstract philosophy or metaphysics. It's about the very real delay between when a stimulus touches your system and when you *actually* feel, see, know, or act on it.

That delay has a name in UED: ATT – Alignment Threshold Time. And the moment ATT is met, reality emerges—not before.

We'll unpack how perception, memory, intuition, and even sudden realizations are not magical powers, but precise ripple events waiting for the right alignment.

Once you see it, you'll understand: You don't just wake up each morning—you wake up every moment your mind aligns.

Consciousness – Not a Light Switch, But a Ripple Window

Here we will redefine consciousness not as a static, always-on state, but as a dynamic ripple window that opens and closes depending on alignment. Consciousness is not the light—we are not bulbs. It's the moment the circuit completes, when the ripple fits.

We often speak of consciousness as though it's binary—you're either awake or asleep, present or absent, aware or unconscious. But under UED, this is not only simplistic, it's misleading.

Consciousness is not a steady flame. It flickers. It pulses. It syncs.

It is not something you *have*—it is something that *happens* when your internal ripple state aligns with the external field. And that alignment doesn't occur continuously. It occurs in windows—brief spans where the Ripple Alignment Criterion (RAC) is met and ATT permits the emergence of experience.

These ripple windows are not tied to external events. You can sit in front of a brilliant sunrise and feel nothing. Or you can suddenly burst into tears recalling a long-forgotten face. The field doesn't care about time. It cares about fit.

Consciousness as a "Perception-Ready" Interval

Think of consciousness as a narrow band of ripple resonance—like a moment when your system clears its static long enough for one clean frequency to cut through. That's when you become aware—not of the world, but of the meeting point between you and the world.

You're Not Always Conscious—You're Occasionally Aligned

Sleep is not unconsciousness—it's simply closed windows. Meditation doesn't create consciousness—it widens the window. Trauma doesn't erase awareness—it clutters the field with unresolved ripples, preventing alignment.

We often mistake presence for being physically awake or alert. But UED reframes presence as attuned availability—an open ripple port where experience may land. When people say, "I wasn't myself," what they truly mean is: "My field wasn't in phase. Nothing could pass through."

Consciousness Fluctuates with Field Conditions

This explains why your awareness surges during danger or silence, why time slows in peak moments, and why some memories vanish only to reappear decades later—because consciousness never records everything, it only records what aligned.

Insight: *"You're not always conscious—you're just occasionally aligned."*

ATT – The Delay Between Impact and Insight

Let me explain how ATT (Alignment Threshold Time) is the fundamental delay between an external event and your internal realization or experience of it. ATT governs when meaning *can* arise—not just when things *happen*.

We often confuse reaction time with awareness. But the brain is not a stopwatch. You don't perceive reality the moment it occurs—you perceive it when the field lets you.

This delay, under UED, is not a bug. It's not slowness. It's a timing gate called Alignment Threshold Time (ATT)—the minimum time needed for a ripple consequence to reach clarity through alignment.

You Don't Feel the Now—You Feel the Aligned

When something happens—a sound, a shock, a word—you don't *experience it directly*. You experience it only once your ripple field synchronizes with the impact. Until then, it's just noise.

That's why you often say:

- "I didn't realize what she meant until later."
- "The meaning of that sentence hit me days after reading it."
- "I froze in the moment, but it all came back afterward."

These are ATT delays—not failures of cognition, but ripples waiting for alignment before consequence could emerge.

ATT Is Not Reaction Time—It's Ripple Resolution Time

In classical neuroscience, we often talk about milliseconds of neural transmission. But in UED, this physical activity is secondary. What matters is when the field permits meaningful resolution—when your internal ripple logic stabilizes enough to form clarity.

For example:

- A trauma event may happen in seconds, but clarity about it might take years.
- A piece of advice may be heard today but understood next year.
- An insult may cause a reaction instantly, but its true emotional meaning lands later—when your field is quiet enough to process it.

This isn't memory delay. It's perception waiting for fit.

Hindsight Isn't Late—It's Field-Matured

ATT explains hindsight—why clarity often blooms only after the fact. Your field wasn't misfiring before; it just wasn't aligned enough yet. Clarity doesn't always come from more facts. It comes from better field conditions.

Insight: *"Meaning never arrives instantly. It settles when the ripple does."*

Why You Notice Some Things—and Not Others

I will show you how perception isn't determined by how *strong* a stimulus is, but by whether it meets the Ripple Alignment Criterion (RAC) and the Alignment Threshold Time (ATT). In other words, *clarity isn't caused by input—it's permitted by fit.*

You and a friend walk into the same room.

You hear a baby crying softly in the next hallway; they don't. You notice a flickering light in the corner; they're focused on the smell of coffee. You both receive identical inputs—but your experiences differ.

This is not attention in the classical sense. It's selective ripple resolution. Your awareness didn't *miss* the other signals—it just didn't *accept* them. Why?

Because your field wasn't aligned to them.

Alignment, Not Intensity, Determines Awareness

It's tempting to think that stronger stimuli win our attention. Loud noise = more awareness. Bright light = more visibility. But that's not always true.

- A whisper can shake you if it resonates.
- A shout can be ignored if your field rejects it.

This isn't a flaw in your attention span—it's the ripple field at work. Everything around you is a field event, trying to resolve through your system. But unless it meets RAC and crosses ATT, it simply doesn't emerge as perception.

Background Isn't Silent—It's Just Unaligned

Much of what we call "background noise" is not unnoticed because it's irrelevant—it's unfit for perception. The field registers everything, but the mind only receives what resolves.

- That clock ticking? You didn't hear it until I mentioned it.
- That hum of a refrigerator? It was there all along, but your field filtered it out—not cognitively, but energetically.

Your mind is not ignoring—it's waiting for resonance.

RAC = Who You Are Right Now

What you notice reveals more about your inner ripple condition than the outer world.

- A hungry person notices food smells.
- A lonely person overhears emotional tones.
- A worried person hears threats in casual speech.

Your field aligns to what it's ready for, not what's simply available.

This also explains why the same person might notice different things at different times. You're not inconsistent—you're dynamically aligned.

Insight: *"You don't filter the world. Your field does."*

Layered Awareness – Fractal Windows of Clarity

Let me demonstrate how awareness is not monolithic but structured in layers, each defined by ATT (Alignment Threshold Time) and ripple coherence, ranging from reflexive sensing to introspective insight.

Consciousness is often treated as a binary: on or off, awake or asleep, aware or unaware. But under Universal Energy Dynamics (UED), consciousness is not a switch—it's a spectrum. A layered phenomenon shaped by ripple alignment and readiness.

Think of it as a ripple ladder. Each step is a *window*—not into the outside world, but into the *clarity* you can handle.

Subconscious: Fast, Low-Resolution, Field-Driven

This is your base layer of awareness. It doesn't think—it *reacts* through stored ripple residue and pattern repetition. Its ATT is extremely short—microseconds.

- You flinch at sudden motion before "knowing" what moved.
- You detect the tone of someone's voice before grasping the meaning.
- You pause on a word you've heard before, even if you don't know why.

This layer is your field on autopilot—fast, instinctive, and largely *unexplained* by your conscious mind.

Conscious Mind: Emotional + Semantic Layer

At this layer, you interpret ripples through language, emotion, and logic. It's where thoughts form, emotions rise, and inner dialogue begins.

- Here, ATT is slower than the subconscious—milliseconds to seconds.
- This layer builds meaning from ripple interactions.
- Emotion arises when alignment is attempted but not reached—like partial resonance.
- Language emerges when patterned ripple states stabilize into repeatable meaning.

It's the realm of speech, debate, decisions—and also confusion, because multiple ripple inputs often fight for coherence here.

Meta-awareness: Introspective, Long-ATT States

The top layer isn't active all the time. It requires long ATT—a window of stillness, reflection, and deeper ripple synchronization. This is where insight, intuition, and self-awareness live.

- It doesn't respond to events—it reviews them.
- It's the layer where you realize something, not just notice it.
- It often arrives late, like understanding a dream after waking, or discovering a truth about yourself months later.

This is not higher because it's better—it's higher because it's broader in ripple span, able to hold contradictions, patterns, and non-linear meaning.

Fractal Windows, Not Isolated Layers

These layers don't operate in silos. They interlock like fractals—each with a nested version of the other.

- A subconscious reflex might trigger a conscious thought.
- A conscious decision might unlock a deep realization later.
- A meta-awareness insight might reprogram subconscious reflexes.

This recursive ripple interaction is what makes human cognition elastic, emotional, and evolving—far more than a stimulus-response machine.

Insight: *"Consciousness isn't one thing. It's a ripple ladder."*

The Moment You 'Get It' – Perception as Completion

Now we will reframe insight, realization, and understanding not as spontaneous bursts of brilliance, but as the natural endpoint of a ripple alignment cycle—a predictable consequence when the field is finally ready.

We often talk about the "aha!" moment as if it were a flash of lightning—sudden, divine, inexplicable. But under Universal Energy Dynamics (UED), this spark is not random. It's the final ripple landing after cycles of partial alignment.

Perception isn't the beginning of understanding. It's the completion.

ATT + RAC = "I Got It"

When you finally "get it," it means two things have occurred:

- ATT (Alignment Threshold Time) has passed: the field has matured to allow consequence.
- RAC (Ripple Alignment Criterion) has stabilized: phase, amplitude, and delay have synced well enough to register a full response.

You don't perceive something when it *arrives*. You perceive it when you align to receive it.

The idea, emotion, or realization was already oscillating in your ripple field. But until your internal state fit the required resonance, it remained unresolved—an incomplete signal, a missing note.

Why Insight Is Delayed, Not Delivered

Often, your mind works in the background, subconsciously juggling ripples, trying different fits.

Then—*click.* e wave locks in. The phase aligns. The amplitude is clear. The time is right.

Suddenly:

- A confusing lecture makes sense.
- A complex emotion simplifies into a word.
- A lifelong belief gets updated in a heartbeat.

It wasn't new information that made it possible. It was completed alignment.

Memory Is Retained Coherence

What you remember long-term isn't what you rehearsed the most. It's what ripple-aligned most completely.

- Emotionally intense moments stay because they formed deep ripple grooves.
- Ideas that "clicked" once are easier to recall—they had high coherence on imprint.
- Trauma lives not because of duration, but because of violent or repeated misalignment that locked in place.

Learning, healing, evolving—they all rely on the stability and replay of that aligned ripple condition.

Peak Awareness = Field Synchrony

Consciousness spikes not when we see more, but when what we see fits completely.

That's why:

- Meditation increases awareness—not by adding input, but by reducing ripple noise.
- Flow states in music, sports, or art feel divine—because RAC holds for extended time.
- Epiphanies strike in silence, in showers, or in sleep—moments

when interference drops low enough for clarity to form.

The "aha" is not information gain. It is field coherence—the moment your mind, memory, and environment stop arguing and start harmonizing.

Insight: *"You don't become conscious of something. You become coherent with it."*

Awareness Is Ripple Timing

Consciousness has long been framed as either "on" or "off," as though it were a light switch flipped by biology or intention. But under the UED framework, consciousness is neither binary nor passive—it is a dynamic ripple process, governed entirely by timing and alignment.

Consciousness is Not a State, But a Readiness

You're not always "aware." You're sometimes ready to align.

Awareness is the window, not the light. It opens when the field conditions permit—not simply when something happens.

ATT Defines When Perception Is Allowed

ATT (Alignment Threshold Time) is the delay between ripple impact and meaningful realization.

It explains:

- Why realization often comes *after* events.
- Why you "sleep on it" and wake up with clarity.
- Why rushed thinking rarely produces wisdom.

You do not perceive at the moment of contact, but at the moment of field permission.

RAC Determines What Is Perceived

Even when ATT is satisfied, RAC (Ripple Alignment Criterion) decides whether the information is fit for resolution:

- If the phase is off → distortion
- If the amplitude is weak → vague awareness

- If the delay is too great → missed or forgotten

This is why two people in the same situation perceive entirely different things—not because of knowledge, but because of field difference.

You Are Not the Light—You Are the Window

Your mind doesn't generate awareness. It permits it.

You are a lens—not the source of truth, but the gateway through which truth becomes visible.

Sometimes the window is clear. Sometimes it's fogged by emotion, memory, fatigue, or interference. But it's always the ripple that determines what gets through.

Closing Line:

"Consciousness is not presence—it's permission. You wake up not to the world, but to the ripple that fits."

Interlude: You Were Always in the Experiment

We began with physics. Slits, photons, interference, tunneling—Elegant mysteries that science boxed in with probabilities and collapse.

But somewhere along the way, the slit turned into your eyes. The pattern became your thoughts. And the interference? That was your emotion whispering, *"I don't fit here."*

The double slit was never just a lab trick. It was a mirror—Reflecting how you see, feel, decide, and sometimes fall apart.

You learned that your vision isn't constant. It waits for permission from the field. That stress isn't pressure—it's phase distortion. That your thoughts don't lag because you're broken, But because the ripple hasn't settled yet.

You discovered that your consciousness is not a flashlight, but a window— Opening and closing based on ATT (Alignment Threshold Time) and filtered by RAC (Ripple Alignment Criterion).

And in that realization, the observer changed.

You are not a passive witness of reality. You are the tuner—A sentient alignment process experiencing clarity or confusion not as a matter of will, but as a function of ripple trust.

You were never just watching the experiment. You *are* the experiment.

And the universe? It's not hiding from you. It's just waiting to fit.

Now, let's go deeper. It's time to question the next great illusion. What if free will...was just ripple behavior you hadn't read yet?

Chapter 14: When the Universe Is Aligning – Stay Out

You are not always meant to act. Sometimes, the best thing the observer can do... is observe.

We've been trained to treat uncertainty like a problem—an enemy to out-think, out-run, or out-work. But under UED, uncertainty is not failure. It's signal. It's the ripple field recalibrating—quietly, carefully—so something precise, meaningful, and consequential can emerge.

In the classical world, action is glorified. In the ripple world, timing is sacred.

Just like a wave forming at sea, you don't splash louder to accelerate it. You wait for its readiness—for Alignment Threshold Time (ATT) to cross, for the field to give you the green light.

This chapter isn't about doing more. It's about understanding why the universe sometimes pauses you on purpose—because you're too close to interfere, and too important to mess it up.

Sometimes the silence isn't empty. It's alignment whispering: "Wait. I'm not done."

The Illusion of Action in Chaos

Let's challenge the modern reflex to interfere when things feel uncertain.

We've been taught to act—to chase clarity the moment confusion creeps in. Culturally, we celebrate intervention: doing something, anything, to force resolution. Uncertainty is demonized. Delay feels like weakness. And so, we push. We scramble. We "take control."

But in ripple dynamics, this reflex is a trap. The structured vacuum doesn't respond to noise. It responds to alignment. And alignment isn't forced—it matures.

When we perceive chaos, we assume disorder. But UED reminds us: what we call "chaos" is often premature observation. The field hasn't completed its recalibration. The ripples haven't met ATT. RAC hasn't stabilized. The outcome isn't ready—because the universe is still thinking.

And here lies the paradox: The more we act in this sensitive zone, the more we distort the very field we want to stabilize. Like shouting into an echo chamber, our panic amplifies the noise—and collapses the alignment we could've patiently received.

Take a common example: You're working through a personal crisis. Uncertainty looms. You don't know what's right, what to feel, or what to say. And instead of letting the field settle, you force conversations, take impulsive decisions, or demand answers. And what happens? More confusion. More ripple noise. More misalignment.

Why?

Because ego wants closure. But the field—it wants coherence.

In ripple logic, action before alignment is interference. It's trying to sculpt a wave that hasn't crested yet. It's pressing keys before the software boots. It's asking for answers before the question is even formed.

This is why mystics were silent. Why sages waited. Why nature never hurries, but everything still gets done.

Insight: *"Not all silence needs noise. Some silences are alignment in progress."*

Field Recalibration – What You Don't See Working

Let's reveal that uncertainty often means the field is reorganizing.

When nothing seems to be happening, everything might actually be happening—just not in the way you're trained to perceive.

In a world obsessed with visible action, silence feels like failure. We equate motion with progress and results with control. But within Universal Energy Dynamics (UED), the absence of surface movement is not absence of activity—it's field recalibration.

Ripple fields are not static. Even when no outcome appears, the consequence vectors are rebalancing beneath your awareness. Tensions are realigning. Noise is dissipating. The system is rediscovering coherence.

But this process is quiet. Unseen. Not because it lacks power, but because it doesn't need display.

Think of emotional healing. The tears are not the repair. The silence after them is. The fatigue isn't the failure—it's your field pausing to restore alignment. Or take physical injury: your body doesn't heal under the surgeon's scalpel alone—it heals in the recovery room, in the sleep, in the slow, unglamorous recalibration of tissue and chemistry. The same is true of the field.

Real change always feels like nothing—until it completes. Just like a ripple doesn't announce its journey—it simply emerges as consequence when the conditions fit.

That's why breakthroughs often arrive suddenly after a long pause. Not because they were spontaneous. But because the field finally aligned. Think of moments where a solution "just came to you" while you were doing nothing. That's not magic. That's ripple logic finishing its background process.

So what do you do during these recalibrations?

You wait. You tune down. You trust. Trying to 'speed it up' is like shaking a compass to get direction. Clarity requires stillness.

Insight: *"The field doesn't need your noise—it needs space to listen to itself."*

ATT as Cosmic Patience

Let's redefine patience not as waiting passively, but as allowing ATT to complete.

In a results-obsessed culture, patience is often mistaken for weakness. We've been taught to "act fast," "decide now," and "push forward." But nature—the true executor of consequence—does not rush. It aligns. And alignment takes time.

In the language of UED, that time is ATT: Alignment Threshold Time—the natural, necessary delay between intent and clarity.

Impatience is nothing but the ego trying to perceive before the field is ready. It's your mind demanding a picture before the lens has focused. It's premature curiosity—not readiness.

You don't see a flower bloom faster by pulling its petals. You don't gain insight by shouting into the void of confusion. And you don't make reality cooperate by stepping in before it's aligned.

ATT is not a flaw. It's a design. The ripple field doesn't obey urgency—it follows synchronization. Every act of true understanding, healing, success, or realization has one thing in common: it emerged when the time was right, not when pressure was highest.

Most failures aren't because someone was wrong. They're because someone moved before the field gave permission.

- Launching an idea before the audience was aligned.
- Speaking a truth before the other could receive it.
- Trusting someone before your own inner ripple reached clarity.

Every one of those is a misfire—not of will, but of timing.

ATT teaches us this: Real vision doesn't race—it waits for the ripple to reach you. It's not a delay in reality. It's a delay in fit.

So the next time you're stuck in ambiguity, tension, or fog...Don't reach for control. Don't manufacture a decision. Just ask: "Has ATT passed? Has the field whispered readiness?"

Because you don't align life by demanding outcome—you align it by allowing emergence.

Insight: *"When consequence hasn't landed, let the ripple arrive first."*

Let It Align – The Wisdom of Inactivity

Now let me explain when withdrawal is the highest form of participation.

We live in a world where doing is the default. Silence is awkward. Pause feels like failure. And inaction is misjudged as weakness or surrender.

But what if stillness isn't passive? What if it's the most ripple-aware decision you can make?

In UED terms, not acting can often be the most aligned response—because the field may still be resolving tension vectors. Your forced move can interrupt a process that's already realigning on its own. In other words: you don't always help things align by interfering. Sometimes, you only unmake what was about to become.

Let's take a practical example.

A heated argument with someone you care about. They're not hearing you. You're not hearing them. Voices rise, logic dies, and consequence is hijacked by emotion.

What's the most powerful thing to do?

Step back. Withdraw. Not to avoid—but to stop misalignment from multiplying.

Just like a ripple cannot form coherent patterns amidst noise, your clarity cannot land while the field is distorted. Withdrawal becomes an active alignment strategy, not a retreat.

This is not just emotional. It's strategic. It's universal. In relationships. In career moves. In decision-making. Even in geopolitical strategy.

The discipline to say, "Let me wait," is not indecisiveness. It's respect for the field's intelligence. It's allowing the invisible processes of alignment to complete—so your next step doesn't collide with unfinished ripples.

Think of the greatest turns in history, in healing, in transformation. They didn't happen in chaos. They happened after it settled. And more often than not, the one who stepped back at the right moment became the one who stepped forward with precision.

- You don't always have to explain.
- You don't always have to push.
- You don't always have to act.

Sometimes, the highest participation is the absence of resistance.

Insight: *"You don't always make things happen. Sometimes, you stop unmaking them."*

The Role of the Witness in Ripple Philosophy

Let's offer the role of the observer as a trusted, non-intrusive presence.

In a world obsessed with engineering every result, *witnessing* feels powerless. But under Universal Energy Dynamics (UED), witnessing is not weakness—it is wise alignment.

Mainstream physics once placed the observer at the edge of phenomena—as if merely *watching* the universe causes it to collapse into form. But collapse is a myth. And control is an illusion.

In ripple philosophy, you don't force the field—you trust it.

Ripples align in trust, not under surveillance. Your presence isn't to dictate; it's to offer coherence—to *be available* without interference. This is the quiet power of the non-intrusive observer.

Just like a calm lake reveals the moon's reflection only when untouched, the field reveals consequence only to the witness who doesn't disturb it.

Consider these moments:

- A therapist sitting silently while a patient finds their own insight
- A teacher allowing the student to struggle through the solution instead of solving it for them

- A parent waiting for their child to return—not with punishment, but with presence

These are not voids. These are vessels of alignment. You become a field—not a force. A partner to possibility—not a manipulator of outcome.

In UED terms:

- You're not the one pushing the ripple
- You're not even defining the ripple
- You are holding space for it to stabilize

This transforms how we live, love, lead, and learn. We step out of the ego's compulsion to *do* and into the ripple's invitation to *be*.

The ripple field doesn't need engineers—it needs resonators. People who *understand when not to interfere*, when to pause, when to let the alignment mature on its own. Because when you try to build a bridge before the river settles, all you make is debris.

Insight: *"You are not the engineer of outcome. You are the partner of alignment."*

Let the Field Finish

We often mistake silence for emptiness, and delay for disorder. But under the lens of Universal Energy Dynamics (UED), these are not signs of absence—they are signals of active calibration.

The chapter redefines uncertainty not as failure or gap, but as a sacred pause—where ripple alignment is still settling. Much like the stillness before a storm or the inhale before speech, this interval is not void—it's value in formation.

- When things seem stuck or unclear, it doesn't mean the universe has stalled. It means the field is realigning—without your interference.
- Field recalibration is a process of balancing ripple tensions,

restoring symmetry, and negotiating consequence. It can't be rushed or forced.

- ATT (Alignment Threshold Time) sets the rhythm. RAC (Ripple Alignment Criterion) sets the permission. You can't override either with intention alone. Your role is often to witness, not wield.

This isn't passivity. It's mastery in stillness. In fact, many of the worst mistakes we make come not from doing too little, but from intervening too early.

We've seen how:

- Pushing through uncertainty increases misalignment
- Real change begins invisibly before it emerges outwardly
- Cosmic patience is not denial—it's deeper intelligence at work
- Inaction isn't failure—it's alignment choosing its own timing

Trust isn't about hope—it's about understanding how the ripple works. When you internalize this, your anxiety transforms into alignment. Your restlessness becomes presence.

Closing Line: "When the ripple is at work, silence is wisdom. The field was never late—it was preparing you to fit."

Chapter 15: Rethinking Randomness

We often say, "It just happened randomly." A storm out of nowhere. A lost wallet. A sudden breakup. A surprise opportunity. Randomness has become our intellectual shortcut—a polite way of saying, "I don't know why this happened."

But what if randomness isn't real?

What if the universe doesn't roll dice—it aligns ripples? What if every so-called random event was just a consequence whose alignment you couldn't see yet? Just because something doesn't make sense to you doesn't mean it didn't make sense to the field.

Universal Energy Dynamics reframes randomness not as disorder, but as invisible choreography. What you call chance is ripple tension resolving quietly—without your approval, and often without your awareness. It's not a glitch in reality. It's just a ripple that didn't ask for your attention.

This chapter challenges one of the most cherished myths of modern science and daily life—that some things "just happen." We'll explore why that illusion arises, how ATT (Alignment Threshold Time) explains the gap between cause and consequence, and why the universe is far more precise than we're patient enough to perceive.

Welcome to a world where nothing is random—only unread.

The Myth of Randomness

Let's challenge the foundational belief that randomness governs certain events in nature and mind.

Randomness is one of the most culturally accepted illusions in modern thought. It exists not because nature behaves erratically, but because we created a term to fill in for moments when our perception couldn't keep up with the process. When things don't follow our expected path, we don't say, "We missed something." We say, "It's random." Convenient. Comforting. Inaccurate.

Let's look at its origins. Early humans used randomness to make sense of the unpredictable—weather, illness, love, death. Over time, we wrapped the idea in fancier terms: chance, chaos, probability, entropy. In quantum mechanics, we elevated randomness to a principle—declaring that at the most fundamental level, nature is inherently indeterminate. It's as if the universe, for all its intelligence, occasionally just flips a coin.

But Universal Energy Dynamics (UED) reframes this assumption. There is no such thing as an uncaused event. There are only events we are unprepared—or unwilling—to align with. Ripple fields don't behave randomly; they behave consequentially. Every ripple interaction is shaped by memory, coherence, tension, and threshold. If a particular outcome seems disconnected from any cause, it simply means we're not seeing the field in full. The consequence was aligned—but not with *us*.

Think of a dice roll. Traditional logic says the outcome is random. But from a ripple standpoint, the throw, angle, grip, air resistance, and table surface all interact as a multi-ripple equation. The result is inevitable—just too complex for us to track. We invented "chance" to excuse our blindness.

In relationships, someone's sudden departure feels random—but it's not. It's ripple misalignment matured into consequence. In quantum physics, the spontaneous emission of a photon isn't random—it's a delayed alignment event based on ATT and field readiness. What's random to a mind stuck in time is rhythm to a field that remembers everything.

Randomness, then, is not a property of nature. It's a label we slap on top of consequences we didn't map.

Insight: *"What we call random is simply what we didn't wait long enough to understand."*

Consequence Has a Pattern—Even If You Can't See It

Here we will show that every ripple-based consequence follows alignment laws, even if its cause is non-obvious.

The universe is not playing dice—it's just playing deep. What looks like chaos from one lens is harmony seen too narrowly. The ripple field, in the Universal Energy Dynamics (UED) framework, stores not only movement but *memory*—it carries the record of past alignments, tensions, and residual interactions. What you see as a sudden consequence is actually the tail-end of a long, silent conversation between forces.

Let's make this personal.

You walk into a room and feel a strange discomfort. Nothing's wrong, but you sense it. A friend walks in moments later, visibly upset. You wonder—did I "sense" their mood before they arrived? Was it random intuition? Coincidence?

No. It was field alignment in action. You aligned with the emotional ripple that was already in play—long before your friend appeared. The field had stored their emotional tension; your subconscious RAC (Ripple Alignment Criterion) picked it up. Coincidence isn't chaos—it's convergence.

This same ripple alignment logic explains much of what we mislabel as randomness:

- A person thinks of a friend they haven't spoken to in years, and suddenly that friend calls.
- You miss a train, only to find out later there was an accident on the next route.
- A child stumbles upon a book that shapes their destiny—but no one remembers how it ended up on the shelf.

These aren't flukes. They are ripple alignments forming across layers of consequence. The field has pattern. It always did. You just weren't tracking enough threads.

Modern science often filters for only *visible causality*—what it can count, measure, isolate. UED looks for *consequential connectivity*. That includes invisible tensions, historical ripples, and unmeasured trajectories still lingering in the field. Every consequence has a lineage—even if your instruments can't trace it.

The universe doesn't work on isolated commands. It flows through synchronized permissions.

So when something "unexplainable" happens, don't rush to label it chance. Ask instead: *What alignment did I miss? What history is the field remembering that I forgot?*

Insight: *"The universe isn't unpredictable. It's just subtle."*

ATT – The Time Randomness Waits to Make Sense

Let's reframe ATT (Alignment Threshold Time) as the hidden clock behind unpredictable perception.

Most people think events happen *when* they're seen. But in the UED framework, *seeing* is the last thing that happens. First, the ripple field builds. Then, alignment occurs. And only *then* does perception allow consequence to show up.

This delay between the origin of a ripple and its observable effect is governed by ATT—Alignment Threshold Time. It's the invisible timer that determines when something becomes available to perception, not when it begins.

Let's consider a familiar example: the "sudden insight."

A scientist stares at a problem for days. Nothing. Then one morning—out of nowhere—it clicks. Did the solution appear randomly? No. The field had been rippling with attempts, associations, failed alignments, and dormant memory structures. The click happened only when ATT was met—when enough ripple harmony built up to cross the perceptual threshold.

The same principle holds in real life:

- A startup struggles for years—then "suddenly" goes viral.
- A relationship seems broken—then one conversation transforms everything.
- You feel stuck for months—then one walk clears your mind.

These are not random turnarounds. They are *delayed recognitions.* The consequence was preparing, resonating, waiting—but your perception hadn't caught up. You were watching for fireworks while the fuse was still aligning.

In a culture obsessed with immediacy, ATT is easy to ignore. We expect real-time outcomes. We assume anything that takes time is either inefficient or broken. But nature doesn't honor urgency—it honors fit. And fit takes time.

Short attention spans produce false randomness.

If you judge the world by what happens *immediately*, everything looks chaotic. But if you let ATT do its work—if you give the field time to align—you begin to see what others miss: the choreography beneath the chaos.

That's why meditation works. That's why incubation boosts creativity. That's why wisdom isn't loud—it's *timed.* Because the ripple doesn't reveal its outcome until the field is ready to echo it back.

So when you're facing a string of "random" events—pause.

The universe isn't being mysterious. It's being methodical.

Insight: *"Randomness is how delayed consequence introduces itself to impatience."*

Why Control Freaks Fear Randomness

Let's link emotional resistance to randomness with an ego desire to dominate timing.

Let's face it: unpredictability makes people nervous. Especially those who equate safety with control.

You'll hear them say things like:

- "I need to know what's coming next."
- "Uncertainty gives me anxiety."
- "If I can't predict it, how can I trust it?"

What's really happening here?

The ego—our inner narrator of certainty—feels threatened when the ripple field doesn't obey its scripts. It has learned to associate *control* with *order* and *predictability* with *power*. So when something doesn't go as planned, or when outcomes feel "random," the ego panics.

But here's the twist: It's not the event that's uncertain. It's your alignment with it.

In UED, the field doesn't malfunction—it misaligns. Not because it's broken or chaotic, but because it is operating *according to fit*, not *desire*. That's the hard pill to swallow.

Desire says, *"I want this now."* Ripple dynamics say, *"Only when you're ready—and so is the field."*

This disconnect creates friction.

Control freaks—those deeply invested in forecasting, managing, and scripting every detail—feel most betrayed by "randomness." But what they're really experiencing is not betrayal. It's a misunderstanding of timing.

They're trying to force perception before ATT, demand consequence before RAC, and script outcomes in a system that rewards readiness—not pressure.

Ironically, this very attempt to control outcomes increases misalignment:

- Overplanning introduces ripple noise.
- Overthinking blocks natural field resonance.
- Overreacting disturbs subtle harmonies the field is forming.

So what looks like randomness is often the field *protecting* itself from intrusion.

And here's the liberating truth: Letting go is not surrendering to chaos. It's aligning with the rhythm already unfolding.

It's saying: "I trust that the field is doing its job, even if I can't see the full picture yet."

This is not passivity. It's participation through patience.

The alternative is exhausting—living in fear of every delayed outcome, interpreting silence as sabotage, and mistaking misalignment for failure.

But when you drop the illusion of control, you gain something better: Responsiveness without reactivity. Presence without panic. Flow without fear.

Because the opposite of randomness isn't control. It's trust—in the ripple, in the timing, in the readiness of things to arrive when they must.

Insight: *"The opposite of randomness isn't control—it's trust."*

From Probability to Possibility—The Ripple Way

Let me offer a new mindset to replace statistical thinking with ripple awareness.

Probability is a brilliant invention. It gives us a sense of control over uncertainty by wrapping it in numbers. We roll dice, flip coins, run simulations—assigning percentages to everything from weather forecasts to life insurance to love.

But here's the hidden catch: Probability is not insight—it's institutionalized ignorance.

It doesn't tell you what *will* happen. It tells you what *might*, based on how little you truly know.

When you say something has a 50% chance, you're not predicting reality. You're declaring a lack of alignment—an inability to perceive the ripple consequences in motion.

That's the core distinction UED brings:

- Probability = guessing when and where alignment might happen.
- Ripple awareness = sensing whether alignment is currently *possible* or *not*.

You don't need to know the odds. You need to read the field.

In ripple logic:

- A "low probability" event can still occur—if alignment is met.
- A "high probability" event can still fail—if readiness is off.

This redefines how we think about possibilities—not as statistical projections, but as consequence potentials waiting for fit.

Take life decisions:

- You don't marry because it's statistically time.
- You don't start a business because trends say it's safe.
- You don't speak because odds favor your point being accepted.

You do these things when clarity appears—when the field is open, the intent is stable, and the ripple supports emergence.

This means moving from:

- Forecasting to feeling fit
- Planning probability to perceiving possibility
- Risk management to resonance mapping

That's not being irrational. That's being *real.*

Because in reality, the field isn't playing games of chance. It's negotiating countless ripple interactions—quietly, precisely, and without favoritism.

And the more attuned you are, the less you rely on statistics and the more you act from alignment.

Insight: *"Don't play the odds—play the alignment."*

It's not about how many times something could happen. It's about whether now is the moment it *should.*

Random Is Just Unseen Order

Randomness has long been a refuge for human confusion. From ancient omens to modern probability theory, we've used the label "random" as a placeholder for what we cannot trace, predict, or explain. But from the ripple perspective in UED, randomness isn't a reality—it's a reflection of our unreadiness.

At its core, randomness is simply perceptual illusion—a mirage that arises when ripple alignments have not completed, or when the observer's own ATT (Alignment Threshold Time) hasn't reached the maturity required for insight. We don't see randomness because it exists—we see it because our field failed to align with the consequence forming underneath.

When something "suddenly" happens—a lucky break, a failure, a flash of intuition—it isn't sudden. It's ripple logic maturing silently beneath the threshold of our awareness, finally reaching a tipping point where RAC (Ripple Alignment Criterion) is met. That's when we *perceive.* Not before. Not beyond.

Here's what we've reframed:

- Randomness is not an event property. It's a symptom of unresolved ripple tension and an impatient observer.
- ATT isn't reaction time. It's the field's rhythm for allowing awareness—a universal patience mechanism.
- RAC is the boundary of vision. It governs not what exists, but *what is allowed to be perceived.*
- Nothing is ever causeless. What seems chaotic is consequence playing out without our permission to see—yet.

UED doesn't invalidate surprise—it explains it. It doesn't remove wonder—it anchors it in coherence.

This isn't an attack on probability. It's an upgrade—from modeling incomplete systems to tuning into a complete field. From playing chance to reading alignment.

So next time life throws you a curveball, don't curse randomness.

Ask: *What did I miss? What ripple matured beyond my window of readiness? Where was the alignment silently forming?*

Because truth is: *"What you called random was never rogue—it was ripple logic, running quietly until the field said yes."*

Chapter 16: The Real Uncertainty Principle

They told you uncertainty is a flaw in nature—a crack in the universe that not even God could smooth over. That some things are just unknowable, untouchable, unmeasurable. But what if they were just... *unfit*?

What if reality never hid? What if it waited? Not for your tools, not for your logic—but for your alignment?

Heisenberg wrote a boundary in chalk. We're carving the real one in ripple.

Because the universe doesn't gamble. It calculates silently until your presence is precise. Until your *Amplitude*, your *Timing*, and your *Phase* match its whisper.

Only then—only then—does the world reveal itself.

Insight: *"Uncertainty is not a cosmic error. It's a cosmic courtesy—waiting until you're ready to see."*

Why Heisenberg Only Got Half the Picture

Let's first contextualize classical uncertainty as a perception-level symptom—not a fundamental law.

Heisenberg's Uncertainty Principle, for decades, stood as a pillar of quantum weirdness. It told us that we cannot simultaneously know a particle's position and momentum with precision. The more we try to pin down one, the fuzzier the other becomes. Why? Because, we were told, the act of measuring *disturbs* the system—like trying to weigh a feather by slapping it with a hammer.

But what if the feather didn't need slapping? What if the disturbance was never the root cause—but merely a symptom?

From the perspective of Universal Energy Dynamics (UED), uncertainty doesn't arise because measurement intrudes. It arises when *ripple alignment Criterions* (RAC) are not met. Reality becomes uncertain to us not because it's hiding, but because we're looking with the wrong rhythm. We're out of sync.

Heisenberg gave us a window into a deeper truth—but he framed it as a limitation. In truth, it's a misalignment.

The inability to measure both position and momentum is not a physical limitation. It's an observer alignment issue. When ATT (Alignment Threshold Time) is insufficient, when amplitude mismatch (A_o / A_r) is too wide, and when phase difference ($\Delta\Phi$) is too erratic, the field refuses perception. It's not that reality breaks down—it simply stops responding.

Just like you can't tune into a radio station unless your receiver matches the frequency, you can't observe reality unless your ripple state matches what the field is ready to share. Heisenberg's disturbance is not the cause of uncertainty—it's a late-stage effect of misfit.

So yes, classical uncertainty *looks* like a hard rule. But under UED, it's simply the echo of unreadiness—a condition of the observer, not the universe.

Insight: *"You don't need to disturb reality to miss it. You just need to misalign."*

Deconstructing the Equation

Now let me explain the components and significance of UED's uncertainty formulation.

In mainstream quantum mechanics, uncertainty is rendered as a mystical boundary—something that restricts how much we can know. UED, however, offers a clean alternative: uncertainty is not a boundary—it's a conditional readiness equation. It's not about what *can't* be known. It's about *when* and *how* it becomes knowable.

The equation:

$$\left(\frac{A_o}{A_r}\right) \cdot \Delta\tau \cdot \cos(\Delta\Phi) \geq ATT_{\min}$$

...is not just mathematical. It's philosophical. It reflects the ripple-centric truth that you only perceive when you're qualified to align.

Let's break it down:

A_0 / A_r – Observer-to-Ripple Amplitude Ratio

This is the clarity or *readiness* of the observer compared to the strength of the ripple signal.

- A_o: Observer amplitude—your current field state, intent, emotional stability, cognitive readiness.
- A_r: Ripple amplitude—signal strength of the external field.
- If $A_o \gg A_r$: You overpower the signal with noise—ego, overthinking, impatience.
- If $A_o \ll A_r$: You're too weak to stabilize perception—fatigue, distraction, incoherence.
- Ideal alignment occurs when this ratio is near 1—perfect resonance.

Δτ – Temporal Delay (ATT Window)

This is the time required to align perception with reality.

- It's not "reaction time" in the neurological sense.
- It's field resolution time—how long your internal ripple state needs to organize itself into perceptual readiness.
- Fast stimuli aren't always useful. In fact, rushing shortens Δτ, reducing clarity.

- Patience becomes physics.

cos(ΔΦ) – Phase Difference (Cognitive Coherence)

This represents the alignment of wave phase between your internal field and the external ripple source.

- **ΔΦ=0:** Perfect coherence. Perception is clear, decisions feel "obvious."
- **ΔΦ=π/2:** Maximum mismatch. You're perceiving noise, not signal.
- The cosine ensures that perception strength scales with mental-phase matching—not just timing or amplitude.

Putting It Together:

All three components must multiply to cross a threshold:

- You may have a strong signal (low A_0 / A_r)
- You may have enough time (high $\Delta\tau$)
- But if you're mentally out of phase (low cos(ΔΦ))—no clarity will occur.

This is RAC in equation form: Perception = amplitude trust × patience × coherence.

The product must exceed:

$$ATT_{\text{min}} = \text{Minimum Alignment Threshold Time}$$

If it doesn't, you don't get to perceive. Not because reality hides—but because you didn't qualify.

Insight: *"You're not allowed to know until the field says you're ready."*

And readiness isn't a wish—it's ripple discipline.

Alignment Is the New Measurement

Now let's replace statistical uncertainty with alignment-based perception.

For over a century, we've treated uncertainty as a limitation of instruments, of knowledge, or of nature itself. We built statistics, probabilities, and tolerances around the idea that *some things simply cannot be measured precisely*. But what if that entire paradigm is misframed?

Under UED, uncertainty is not the failure of instruments—it's the failure of alignment. The world isn't hiding. You're just not tuned.

Measurement Is Not About Data—It's About Field Fit

In classical science, a "measurement" is treated as a captured value: something taken, stored, and verified. But in ripple logic, measurement is alignment:

- No alignment → no perception.
- No perception → no measurement.
- Measurement is simply the event of awareness in action.

If you didn't measure something, it's not because it was probabilistically blurry—it's because the field didn't permit you to see it.

Why Instruments Still Mislead

Scientific instruments are often treated as impartial observers. But in truth:

- Every device has its own amplitude ratio (A_o / A_r), timing window $\Delta\tau$, and internal phase noise $\Delta\Phi$.
- They are not immune to misalignment. They simply offer a more stable RAC—not a perfect one.
- Thus, even our tools are ripple participants, not rulers of reality.

ATT_{min} – The Hidden Speed Limit of Awareness

The UED uncertainty equation:

$$\left(\frac{A_o}{A_r}\right) \cdot \Delta\tau \cdot \cos(\Delta\Phi) \geq ATT_{\min}$$

...tells us that unless the total coherence of observer and event crosses a threshold, perception doesn't occur.

- This ATT_{min} isn't just a number—it's a fundamental ripple permission delay.
- It's the real bottleneck behind what we mislabel as random, fuzzy, or "uncertain."

So what appears as statistical scatter is, in truth, a non-aligned field scenario. Once alignment is achieved, what was fuzzy becomes sharp. What was impossible becomes obvious.

Phase, Time, and Amplitude—All or None

- You can't compensate a weak amplitude with more time.
- You can't overcome phase mismatch with signal strength.
- You can't perceive without all three being *simultaneously tuned.*
- Measurement is not a partial privilege—it's a full-field license.

Insight: *"Uncertainty isn't blur—it's timing out of tune."*

The truth is never absent. It's simply unavailable—until the ripple clicks into place.

What This Means for Science, Mind, and Reality

Now we will draw sweeping implications across physics, consciousness, and behavior.

We've built entire civilizations on fragmented metaphors—particles that are waves, observers that disturb, consciousness that glows, randomness that governs. But UED's final alignment principle doesn't just tweak the model. It rewrites the foundation.

This isn't a new theory of uncertainty. It's a new agreement with reality.

No More Wave-Particle Duality

The duality never existed in the field. It existed in our *interpretation.*

- A ripple is neither wave nor particle—it's a field event awaiting alignment.
- When perception lands mid-cycle, we call it a wave.
- When it lands post-collapse, we call it a particle.
- When it never aligns, we call it a paradox.

Under UED, we no longer chase what light *is*—we understand what light *requires*. And that's RAC: Ripple Alignment Criterions.

Consciousness: Not the Light, But the Receiver

We used to think the mind "shines a light" on things to understand them.

But alignment teaches us that consciousness doesn't illuminate—it tunes in.

- It doesn't impose clarity. It waits for the field to allow it.
- Attention isn't a mental act—it's a ripple sync request.
- The *feeling* of knowing is the consequence of coherence, not the cause.

Thus, consciousness isn't proof of awareness. It is the effect of alignment.

Reality Isn't Hidden—It's Protected by Fit

Most scientific theories obsess over access—how to find, capture, measure, predict.

But UED proposes a deeper respect: Reality doesn't respond to interference. It responds to trust.

- ATT is the universe's timeout switch: "You'll know when you're ready."
- Misalignment doesn't mean the truth is unknowable. Just not *yours* yet.
- Even emotion, intuition, and insight obey this law—nothing is random, only *waiting to resonate.*

So when a breakthrough happens, in science or soul, it was never magic. It was matured permission.

From Observation to Resonance

This transforms the role of the scientist, the philosopher, the seeker. You're not a *collector of truth*. You're a resonator with the field.

- The more coherent your intent, the clearer the alignment.
- The more aligned your inquiry, the deeper the truth revealed.
- You don't *uncover* reality—you synchronize with it.

This applies as much to quantum spin as it does to understanding a loved one's silence.

Insight: *"The universe never withholds truth. It just requires fit."*

Truth isn't out there. It's *everywhere*—locked in ripple codes, waiting for the moment when the observer, the question, and the field finally meet in alignment.

The Ripple Observer's Code

Let's offer a practical takeaway: how to live, think, and perceive with the UED principle.

If there's one thing the universe keeps whispering through every ripple, every delay, every silence—it's this:

"Stop trying to dominate what you were meant to align with."

That's the final shift from conventional science to ripple consciousness. From controlling reality to becoming its most precise, perceptive partner.

So what does it actually look like to *live* this philosophy?

1. Tune Your Amplitude – Focus Is Your Strength

In the UED equation, A_o/A_r isn't just math. It's your ratio of personal presence to environmental signal.

- The more scattered your attention, the weaker your observer amplitude.
- The more centered you are, the stronger your field resonance.

Want clarity? Don't demand it. Increase your presence. Want answers? Don't search harder. Listen sharper.

The universe doesn't speak louder than your readiness to receive.

2. Respect Δτ – Don't Rush Insight

The ripple doesn't operate on deadlines.

- Δτ (delta tau) is not a fault. It's a feature of wisdom.
- Rushing it means acting before coherence—like harvesting fruit before it ripens.

Every delay, every pause, every unexplained waiting period is ripples reorganizing. Not punishment—preparation.

So instead of fighting it, ask:

"What's still syncing in the field that I need to wait for?"

3. Align Your Phase – Match Intent With Clarity

The cosine of ΔΦ (phase difference) reflects your internal harmony.

- If your intentions are strong but blurry, you're out of phase.
- If your thoughts are clear but your motive is off, you're out of phase.

You can't fake alignment. You feel it.

True knowing comes when your why, what, and how vibrate in sync.

So phase alignment isn't technical—it's soulful.

4. Wait for ATTmin – Trust the Minimum Time Required

There's always a minimum threshold—a delay nature demands before letting you in.

- ATTmin is your cosmic patience meter.
- It can't be hacked, rushed, or bypassed—not by thought, not by force, not by logic.

The field lets you see when you're aligned *enough*. Until then, stay still.

That idea you're trying to force? That conversation that won't land? That solution that keeps escaping?

Maybe you're not wrong. You're just early.

The Observer's Practice

Living with the Ripple Observer's Code means:

- You tune yourself first.
- You don't chase the outcome—you align to let it land.
- You don't panic at delay—you honor its necessity.

This is not passivity. This is precision.

It's not surrender. It's *sovereignty without sabotage.*

Insight: *"Reality is yours—not by force, but by fitness."*

The universe doesn't deny you. It just waits for you to fit the signal it already holds.

And once you do—no door can remain closed. Because you didn't knock it down. You became the key.

The Last Uncertainty

At the end of all inquiry, what stands between confusion and clarity is not randomness... but readiness.

The UED Uncertainty Principle reframes the fog we call uncertainty—not as a curtain of ignorance, but as a momentary pause in ripple alignment. Nothing is ever truly hidden. Not in the universe. Not in your mind. Not in your emotions.

Only unread.

Let's break it down:

- Misalignment—not indeterminacy—is the real source of unpredictability. Just like a blurry photo isn't a mystery—it's a misfocus.
- The universe is not chaotic; it is subtly ordered. You miss it not because it hides, but because you arrived out of tune.

The UED equation says it all:

$$\left(\frac{A_o}{A_r}\right) \cdot \Delta\tau \cdot \cos(\Delta\Phi) \geq ATT_{\min}$$

Translation?

- A_o/A_r – You must *match your attention to the field's presence.*
- $\Delta\tau$ – You must *allow the delay necessary for alignment to complete.*
- $\cos(\Delta\Phi)$ – You must *be in phase with the intent, emotionally and cognitively.*

Only then are you permitted to know.

Not because the truth changed. But because you did.

This final formulation isn't just a law of perception. It's the unifying grammar of:

- Measurement (You can't know what you aren't ready to observe)
- Emotion (You can't feel what isn't in sync with your field)
- Consciousness (You can't awaken until your ripple is open)
- Entropy (What seems lost is often just unaligned)

Everything—science, soul, self—is governed by this threshold of fit.

So the next time you feel uncertain...

Don't assume ignorance. Don't force a conclusion. Don't label it random.

Ask yourself: *"What part of me needs to realign to read what's already there?"*

Final Closing Line: *"Uncertainty isn't the fog of truth—it's the ripple's pause, waiting for you to be in phase."*

Closing Chapter: The Science of Waiting

"We did not collapse the wave. We misread the ripple." Uncertainty taught us not to calculate more, but to wait better.

Waiting has always been seen as a passive act. In the age of speed, it's often mocked. The faster you think, the quicker you react, the more you're celebrated. "Time is money," we're told. "Act now," we're warned. But when it comes to reality—its most intricate truths, its most intimate signals—the universe doesn't reward haste. It rewards resonance.

Uncertainty was never the problem. Misalignment was.

A New Science Emerges

We've spent centuries trying to measure nature as if it were an open book. But nature is a ripple script—it must be read only when the ink has dried, not while it's still writing itself. From Heisenberg's uncertainty principle to quantum decoherence, we tried to decipher the blur. We mistook that blur for randomness. We mistook probability for principle. We mistook impatience for insight.

Ripple Alignment Criterion (RAC) and Alignment Threshold Time (ATT) flipped the narrative.

They taught us that perception isn't something we extract from reality. It's something we're granted—only when we are in phase, in tune, in trust with the ripple field.

Let that sink in.

You don't learn by digging deeper. You learn by aligning better.

Perception as Permission

For centuries, we believed that observation disrupts reality. That to know something is to shake it loose from its natural state. But what if that's just projection? What if the disturbance isn't caused by light, or instruments, or measurement—but by impatience? By a mind that isn't ready, yet tries anyway?

When your ripple field is unready—out of amplitude, off-phase, or misaligned in time—you will see blur, you will feel randomness, and you will miss meaning. That doesn't make reality broken. It makes it hidden. Not by force, but by ripple ethic.

Perception is granted. Not captured.

You are not just a viewer. You are a variable in the equation of awareness. The clarity of the universe depends on your fit with it.

Why We Misread

Let's go back to that legendary quantum riddle—the double slit.

They said the wave collapsed. That the particle "decided" when we looked. But what if nothing collapsed? What if nothing decided?

What if perception *only occurred* when your ripple field reached sufficient clarity?

The interference pattern wasn't destroyed. You simply didn't align with it.

That's the true science of waiting—not passivity, but field maturity. Let the universe ripple into place. Let the fit form. Stop rushing consequence before it's ready.

Misreadings, from emotion to experiment, stem not from a broken system but a premature observer.

The Observer's Discipline

To know the universe isn't to command it—it's to befriend its rhythm.

Here's what waiting really means under UED:

- You focus not on forcing outcomes, but on fine-tuning your field.

- You don't rush to answer questions, you align to when answers will reveal themselves.
- You release the myth of now or never—because in ripple logic, time doesn't demand reaction. It demands readiness.

This is not a surrender of agency—it's the cultivation of precision. Waiting, in UED, is not delay. It's alignment time.

You wait because the ripple is organizing. You pause because perception is composing. You breathe because consequence is being born.

Alignment Is Everything

The greatest discoveries in history were not "made"—they arrived. When the observer finally aligned. When ATT was complete. When the RAC stabilized. That's when gravity revealed itself. That's when evolution took shape. That's when consciousness stepped into awareness.

It didn't happen *because* someone forced it.

It happened because someone was listening when the ripple finally landed.

Emotion, Mind, Science, and Silence

In every domain—emotion, thought, science, society—the same principle applies. You feel anxious? It's a ripple misfit. You're confused? It's a delay in ATT. You can't make sense of events? It's not chaos—it's recalibration.

Silence isn't absence. It's order forming.

UED doesn't just offer a new physics. It gives us a new psychology. A new spirituality. A new ethics. All anchored in one insight:

Alignment precedes access. If it hasn't arrived yet, it isn't lost. It's still aligning.

What Waiting Gives Us

1. **Confidence in Stillness:** You no longer need to panic in uncertainty. Stillness may be the signal that something

profound is organizing beneath perception.

2. **Clarity in Delay:** The pause between stimulus and meaning is sacred. That is where wisdom grows. Not where understanding is blocked.
3. **Discipline in Perception:** You stop over-analyzing. You start over-aligning.
4. **Respect for Fit:** No more force-fitting the world into premature conclusions. You wait for what fits—because that's what lasts.

The Ripple Path Forward

This book began by questioning what we think we know—about waves, particles, forces, time, light, and even mind. It ends here, in the simplest of truths:

The universe works. You don't need to fix it. You need to fit it.

So next time you're lost, don't rush for answers. Don't force understanding. Don't panic at silence.

Tune your amplitude. Adjust your phase. Hold your patience.

Let the ripple complete. Because when it does, everything makes sense.

Final Insight:

"What you called noise was signal in progress. What you called failure was ATT incomplete. What you called randomness was a ripple waiting to land. You were never alone. You were just early."

Let this be your science of waiting. Not passive. Not blind. But powerful. Precise. And ripple-aligned.

Appendices

Appendix A: Mathematical Derivation of RAC

(Ripple Alignment Criterion)

Purpose:

To formally define and derive the Ripple Alignment Criterion (RAC)—the fundamental perceptual threshold under the Universal Energy Dynamics (UED) framework. RAC determines when a ripple-based event becomes cognitively observable, emotionally perceivable, or scientifically measurable by the observer.

Step 1: Understanding the Observer–Ripple System

In UED, perception is not a passive recording—it is a conditional *alignment event* between the observer's cognitive field and the ripple field of reality. For such perception to take place, three major parameters must satisfy a threshold equation.

Let's define them:

- A_o = Observer's ripple amplitude (internal readiness, attention, focus)
- A_r = Ripple amplitude of the external field (signal strength, stimulus intensity)
- $\Delta\tau$ = Temporal delay between stimulus and observer readiness (Alignment Threshold Time)
- $\Delta\Phi$ = Phase difference between the observer's ripple and the incoming field ripple

These three together govern whether and when the ripple will become observable.

Step 2: The Alignment Product

RAC postulates that perception occurs only when the product of these three alignment terms crosses a critical threshold, defined as:

$$\left(\frac{A_o}{A_r}\right) \cdot \Delta\tau \cdot \cos(\Delta\Phi) \geq ATT_{\min}$$

Where:

- A_o/A_r defines *amplitude coherence* (Too small: observer is unfocused; too large: field is too weak)
- $\Delta\tau$ defines *temporal overlap* (If attention and event are not temporally matched, perception fails)
- $\cos(\Delta\Phi)$ defines *phase alignment* (Out-of-phase perception causes interference, not clarity)
- ATT_{min} is the minimum alignment time required for stable perception to occur.

Step 3: What the Equation Implies

Let's analyze each variable's contribution:

Amplitude Match (A_o/A_r)

- High observer amplitude means high attention or focus.
- But even a strong external ripple won't be seen if attention (A_o) is low.
- A perfect signal (A_r) cannot compensate for a distracted observer.

Time Delay ($\Delta\tau$)

- This is not raw reaction time—it's the time it takes for cognitive or emotional readiness to align with the stimulus.
- If too short, there is no room for internal organization.
- If too long, the ripple disperses before alignment completes.

Phase Alignment ($\cos(\Delta\Phi)$)

- Perfect alignment is $\Delta\Phi = 0$, i.e., $\cos(0)=1$
- Any shift in phase decreases clarity.
- $\Delta\Phi$ approaching 90° ($\cos(\pi/2)=0$) results in total perceptual null.

Step 4: Special Cases and Limits

Case 1: Perfect Alignment

$$\Delta\Phi = 0 \quad \text{and} \quad A_o = A_r \quad \Rightarrow \left(\frac{A_o}{A_r}\right) = 1, \quad \cos(\Delta\Phi) = 1$$

Then:

$$\Delta\tau \geq ATT_{\min}$$

→ Observer must at least wait ATTmin for full perceptual alignment.

Case 2: No Phase Match

$\cos(\Delta\Phi) = 0 \Rightarrow$ Perception impossible, even with perfect amplitude and delay

Case 3: Strong Observer Focus but Weak Signal

$A_r \ll A_o \Rightarrow$ Exaggerated internal readiness with minimal input

→ Overinterpretation, illusion, or false perception—hallucinations.

Step 5: Interpreting ATT_{min} (Threshold of Awareness)

- ATT_{min} represents the lowest possible time-energy coherence required for ripple-based perception to be *valid.*

- In cognitive terms: the time needed for a stimulus to form a complete thought or realization.
- In emotional terms: the delay between feeling and understanding.
- In physical systems: the time required for measurement without interference.

Step 6: From RAC to Reality

This condition is not just a math rule—it governs:

- Why you miss people's words under stress
- Why intuition is fast, but insight is delayed
- Why coincidences appear random
- Why science misreads subtle systems

Conclusion: RAC is the Lens of All Perception

This derivation shows that reality isn't hidden—it's just ripple-coded. RAC reveals that:

- Awareness is a permission granted by the field, not a button pressed by the observer.
- Perception is not mechanical—it is conditional.
- Knowing depends on fit, not force.

$$\text{If } \left(\frac{A_o}{A_r}\right) \cdot \Delta\tau \cdot \cos(\Delta\Phi) \geq ATT_{\min},$$

then you are aligned. You are allowed. You are aware.

Appendix B: UED in a Nutshell

Universal Energy Dynamics Made Simple

Purpose:

To distill the core framework of Universal Energy Dynamics (UED) into a compact, digestible format for reference. This appendix acts as your ripple-map for navigating reality—not through particles, forces, and formulas, but through alignment, consequence, and coherence.

1. The Universe Is Still—Only Ripples Move

In UED, motion is illusion. What you call "movement" is just a shift in ripple alignment. The universe isn't expanding, rotating, or flowing—it's perfectly still. What moves is energy *trying to re-align*.

Insight: "Nothing moves. Everything ripples."

2. Energy = Actioned Intelligence

UED defines energy not as a property, but as a consequence of intelligence seeking resolution.

- Energy is not a thing.
- It's a *manifestation* of universal intent in the field.
- Intelligence acts → ripples arise → we observe energy.

Insight: "Energy doesn't exist by itself. It appears when intelligence acts."

3. Ripple Is the Only True Entity

UED replaces all particles, forces, and wave-particle dualities with one mechanism:

Ripple = Structured consequence seeking equilibrium

- No photon. No electron. No quark. Just ripple-based interactions.
- A ripple has direction, frequency, amplitude, and phase—but most importantly, intention.
- It carries history, meaning, and alignment potential.

4. ATT – Alignment Threshold Time

ATT is the universal timing condition. It governs when:

- An event becomes observable
- A thought becomes clear
- A consequence becomes measurable

If a ripple interaction hasn't crossed ATT, it doesn't exist for the observer—even if it's happening.

Insight: "Nothing is real until it aligns—and waits."

5. RAC – Ripple Alignment Criterion

Perception, memory, learning, emotion, and even scientific observation require:

$$\left(\frac{A_o}{A_r}\right) \cdot \Delta\tau \cdot \cos(\Delta\Phi) \geq ATT_{\min}$$

Where:

- A_o = Observer amplitude (attention)
- A_r = Ripple amplitude (signal strength)
- $\Delta\tau$ = Perceptual delay (ripple wait time)
- $\Delta\Phi$ = Phase difference (cognitive coherence)

Insight: "Awareness is alignment—not effort."

6. Force Doesn't Exist—Only Misalignment Pressure

- What physics calls "force" is just consequence pressure due to field imbalance.
- You don't push a door open—you misalign the field until the consequence resolves.

Insight: "Force is the scream of a ripple trying to find peace."

7. Time Is an Afterthought—Born from ATT

Time is not a dimension. It is a felt consequence of ripple delay. You don't live in time—you live through ripple resolution windows.

- Fast ATT → feels fast
- Slow ATT → feels slow
- No ATT → timelessness (as in sleep, meditation, death)

Insight: "Time doesn't pass. Ripples do."

8. Consciousness Is a Ripple Window

You're not conscious all the time. You're aligned sometimes.

- Perception happens when ripple readiness meets cognitive availability.
- Sleep, intuition, meditation = different ATT zones
- Trauma, distraction, fatigue = RAC failure zones

Insight: "Consciousness is not constant—it's episodic alignment."

9. Suffering = Misalignment

Emotions are ripple echoes.

- Anxiety = Future misalignment
- Sadness = Failed resonance with memory
- Rage = Amplitude overload

- Healing = Realignment

Insight: "You don't suffer because you feel. You suffer because your ripple can't fit."

10. Everything Is Consequence

UED reframes reality from cause-driven to consequence-aware.

- You can't always trace a cause.
- But you can always observe an outcome.
- Consequence = ripple logic unfolding through field intelligence.

Insight: "Cause is a story. Consequence is real."

Final Takeaway: UED Is Not a Theory—It's a Lens

- It doesn't replace physics—it reveals what physics couldn't perceive.
- It doesn't oppose consciousness—it explains how it fits.
- It doesn't fight uncertainty—it clarifies what it waits for.

UED doesn't ask you to believe. It asks you to observe differently.

"You are not a mind in a body, watching the world. You are a ripple in the field—remembering who you were when it finally fits."

Appendix C: UED vs Quantum vs Classical

This section provides a clear, side-by-side comparison of how Universal Energy Dynamics (UED), Quantum Mechanics, and Classical Physics interpret key phenomena. It will help readers contextualize UED as not just a philosophical alternative, but a scientifically structured upgrade—eliminating paradoxes while preserving observable reality.

1. Nature of Reality

Concept	Classical Physics	Quantum Mechanics	Universal Energy Dynamics (UED)
Assumption	Objective, deterministic reality	Probabilistic, observer-influenced	Consequential, ripple-driven
What exists?	Matter and energy in motion	Wavefunctions and probabilities	Still field rippling with structured intelligence
Measurement	Independent of observer	Affects reality (wavefunction collapse)	Alignment-dependent perception—nothing collapses, only aligns

UED Insight: "Reality doesn't collapse. It waits for fit."

2. Observer Role

Concept	Classical	Quantum	UED
Observer	Passive measurer	Active participant	Ripple-based receiver
Effect on system	None	Alters wavefunction	Triggers RAC (Ripple Alignment Criterion)
Awareness	Not considered	Unclear mechanism	Consciousness = alignment, not attention

UED Insight: "You don't observe reality. You ripple into it."

3. Time and Motion

Concept	Classical	Quantum	UED
Time	Absolute (Newton) or relative (Einstein)	A parameter in equations	A perceptual delay (ATT)
Motion	Trajectories of particles	Probability amplitude flows	Field seeking equilibrium through ripple shifts
Cause and Effect	Linear causality	Non-local entanglement	Consequence-first: cause is retro-labelled perception

UED Insight: "Cause is an afterthought. Consequence is real-time."

4. Light and Waves

Concept	Classical	Quantum	UED
Light	Electromagnetic wave (Maxwell)	Photon—quantized particle-wave	Perceived ripple alignment, not a moving object
Wave-Particle Duality	Incompatible dual model	Fundamental paradox (e.g., double slit)	Resolved: Only ripple field. No duality.
Speed of light	Constant	Fundamental limit	Consequence delay, not actual motion

UED Insight: "Light doesn't travel. You align with it."

5. Uncertainty and Measurement

Concept	Classical	Quantum	UED
Uncertainty	Measurement error or limitation	Heisenberg's principle	Alignment threshold not met (ATTmin not reached)
What determines observability?	Instrument precision	Wavefunction collapse	Amplitude match, timing, phase coherence (RAC)

UED Insight: "You didn't miss the data. You missed the alignment."

6. Consciousness and Mind

Concept	Classical	Quantum	UED
Mind role	Ignored	Loosely linked to observation	Central to perception and reality resolution
Consciousness	Not addressed	Speculative	Defined as field-window readiness
Emotion, thought	Biological byproduct	Outside scope	Ripple feedback misalignment or phase delay

UED Insight: "You don't think through neurons. You ripple through resonance."

7. Fundamental Entity

Concept	Classical	Quantum	UED
What's basic?	Mass, energy, force	Fields, particles, uncertainty	Ripple field, structured vacuum, intelligence
Theory of Everything	Elusive	Mathematical unification attempted	Philosophical-physical integration via consequence
Dark matter/energy	Unexplained placeholders	Theoretical extensions	Misread ripple effects from misalignment or invisible RAC

UED Insight: "What's missing isn't matter—it's understanding."

Final Comparison Summary

Criteria	Classical	Quantum	UED
Language	Math and mechanics	Probability and statistics	Ripple logic and consequence coherence
Strength	Predictive models in macro	Accurate in micro, but paradox-prone	Coherent across all scales
Weakness	Can't explain mind, emotion, or randomness	Interpretation confusion, paradoxes	Requires new mindset, reframing observation
Role of Human	Separate from the universe	Intertwined but unexplained	Ripple within the field—co-creator of consequence

UED doesn't dismiss classical or quantum insights. It explains why they work, where they fail, and how to unify both realms—through ripple permission, not particle collision.

Appendix D: Thought Experiments & Simulations

Testing UED with Mind and Model

This appendix presents a curated set of thought experiments and simulation frameworks designed to help readers visualize, question, and test Universal Energy Dynamics (UED) principles—without requiring complex mathematics or machinery. These models bridge conceptual understanding and experiential clarity, allowing both lay readers and scientists to explore ripple logic through simple but profound setups.

1. The Delayed Mirror Test

Purpose: To test Alignment Threshold Time (ATT) and RAC in visual clarity.

Setup: Sit in front of a mirror in a dim room. Dim the light until you can *just barely* see your face. Now breathe slowly, blink gently, and observe your own eyes—watch for the moment when your perception suddenly "clicks" into clarity.

Observation: There is always a perceptible delay or hesitation before recognition becomes solid. This is the ATT—not the light delay, but field readiness aligning your internal state with external input.

UED Insight: Clarity doesn't come with light—it comes when ripple fields allow it.

2. The Emotional Echo Experiment

Purpose: To experience emotional misalignment as ripple feedback.

Setup: Recall a moment when you felt hurt or misunderstood, and journal what you wished had happened instead. Now replay the memory slowly while observing your emotional shifts.

Reflection: You'll notice "pockets" of dissonance—where your memory and emotional response do not match your current understanding. These are RAC disruptions. With each pass, emotional residue may lessen as alignment improves.

UED Insight: Memory isn't static—it ripples until resolved.

3. The Phase Misalignment Test

Purpose: To explore interpersonal misalignment.

Setup: Have two people read a simple script or poem to each other—but with mismatched tones, emotional states, or pace. Then re-read it again, this time in emotional and rhythmic sync.

Observation: The first read may feel off, hollow, or funny. The second—when aligned—feels richer, deeper, and connected. This is amplitude and phase alignment in practice.

UED Insight: You're not just hearing words—you're riding the ripple together.

4. The Attention Window Timer

Purpose: To simulate ATT in cognitive learning.

Setup: Time how long it takes to understand a new, unfamiliar concept (e.g., a poem, quote, or foreign sentence). Record the moment you feel "I got it"—not just reading it, but understanding it.

Measure: This is your ATT duration for that information type. Try again under stress, distraction, or fatigue. The time changes—not because the concept changes, but your field does.

UED Insight: Intelligence doesn't vanish. It delays.

5. The Ripple Drop Simulation (Digital / Physical)

Purpose: To visualize field interactions and RAC visually.

Setup: In water or using ripple simulation software, drop two pebbles with a slight time gap. Observe where the waves intersect constructively (RAC zone) or cancel out (interference zones).

Interpretation: Each intersection is a metaphor for field overlap. Where they harmonize, clarity emerges. Where they diverge, confusion, chaos, or ambiguity results. That's how observation behaves in the universe.

UED Insight: Observation isn't about position—it's about permission through alignment.

6. The Decision Haze Model

Purpose: To simulate how emotional misalignment affects decision-making.

Setup: Take a life decision you're struggling with. On paper, list facts, desired outcomes, and fears. Now rate how clear or hazy each point feels on a 1–10 scale.

Now pause. Meditate or take a walk. Revisit your chart.

You'll find some ratings changed—not because your mind "worked," but because ripples settled and realigned.

UED Insight: Resolution isn't logic-based. It's field-readiness driven.

7. The Randomness Delay Game

Purpose: To internalize ATT's role in so-called "random" events.

Setup: Using a random number generator or dice, guess the outcome—but not instantly. Pause for five seconds. During the pause, try to "feel" the alignment, not guess statistically.

Observation: Many report greater accuracy when pausing. Why? Because you're allowing field resonance—not mental impulse—to guide the perception.

UED Insight: Probability is impatience pretending to be logic.

8. The Ripple Listening Test

Purpose: Explore audio alignment as RAC.

Setup: Play a sound or music at low volume. Sit silently and try to "tune in" before increasing the volume. Focus not on hearing, but feeling the presence of sound.

Result: Often, you'll "sense" the ripple before you audibly hear it. The field meets you first, and the signal follows. This is pre-perception alignment.

UED Insight: Listening starts before ears confirm it.

Final Thought

These experiments aren't just symbolic. They retrain your perception from passive consumption to active ripple engagement. UED isn't a theory to memorize—it's a conscious way to experience reality.

"To prove a ripple exists, you don't have to touch it. You only need to become quiet enough to feel it touch you."

Appendix E: Future Experiments to Validate ATT

(Alignment Threshold Time)

Bridging Perception and Physics through Testable Delay

Universal Energy Dynamics (UED) introduces a radical shift from measuring what exists to perceiving when and how alignment allows something to exist—for the observer. At the heart of this is ATT (Alignment Threshold Time)—the minimal duration required for a ripple-based interaction to be consciously perceived.

To bring this from framework to verifiable science, here are future experimental pathways to validate ATT across cognitive, neurological, physical, and technological domains.

1. Cognitive Resonance Time Trials

Domain: Experimental Psychology / Cognitive Neuroscience

Goal: Identify the minimum time required for a subject to perceive clarity in a pattern.

Setup:

- Flash ambiguous images (e.g., Necker cubes, Rubin vases) to participants for variable milliseconds.
- Measure the moment of recognition versus just visual registration.
- Introduce stress, noise, or emotional context to observe ATT variance.

Expected Finding: Recognition doesn't depend solely on stimulus duration—it depends on when the internal ripple (attention, clarity, emotional state) synchronizes with the input.

Insight: ATT isn't response time. It's field-fit readiness time.

2. Quantum Delay Matching with Human Perception

Domain: Quantum Cognition / Perceptual Physics

Goal: Observe if perceptual "collapse" aligns with quantum delay under conscious attention.

Setup:

- Double-slit photon experiments with conscious vs. unconscious observers (replicating Dean Radin-style trials).
- Instead of outcome collapse, track micro-delay patterns between signal and recognition under different mental states.

Novel Focus: Does human attention shift the timing of wave-interference stabilization, suggesting a field-alignment based effect rather than classic observation collapse?

Insight: The delay between reality and realization is where UED lives.

3. EEG-Based ATT Mapping

Domain: Neuroscience / Brain-Computer Interface

Goal: Establish biometric correlation of ATT using brainwave timing.

Setup:

- Present stimuli at controlled intervals while recording EEG.
- Look for phase-locking patterns (e.g., P300, N400) where recognition or decision-making aligns across trials.
- Introduce attentional shifts to see how ATT modulates under cognitive load.

Output:

- Build an ATT curve across emotional states, distractions, and clarity levels.
- Use AI to model predictive ATT patterns from neural data.

Insight: ATT could become a measurable signature of internal resonance readiness.

4. Phase-Matched Machine Learning Simulations

Domain: AI / Neural Networks / Signal Processing

Goal: Simulate field alignment logic using dynamic perception thresholds.

Setup:

- Train a machine learning model to recognize data only when input matches specific amplitude and phase filters.
- Track how long and under what conditions the system "perceives" correct data.

Why It Matters: This could emulate non-linear alignment time—closer to human intuition—rather than brute-force computation.

Insight: Smart systems need more than speed—they need timed coherence, just like UED perception.

5. Temporal Field Lag in Decision-Making

Domain: Behavioral Science / Decision Theory

Goal: Model how real-life decisions follow ATT thresholds.

Setup:

- Provide incomplete information and watch when participants decide.
- Use response latency, confidence levels, and post-decision analysis to map decision-perception timing.

Test Variables:

- Delay before acting under uncertainty.
- Ripple mismatch (emotional stress, bias, misinformation) that extends ATT.
- Coherent info streams (calm, relevant, emotionally neutral) that shorten ATT.

Insight: Poor decisions aren't irrational—they're often misaligned in time.

6. Cosmological ATT Analogs

Domain: Astrophysics / Observational Physics

Goal: Investigate ripple readiness delays at cosmic scale.

Approach:

- Analyze anomalous light delays, gamma ray bursts, or gravitational lensing events not just by spacetime geometry, but through ripple phase modeling.
- Could cosmic "flickers" or seemingly random bursts be signs of macro-ATT thresholds being crossed?

Speculative Extension:

A "signal" isn't late. It only appears when the observer-field finally fits the event's ripple signature.

Insight: Even starlight may not travel—it may arrive only when alignment completes.

7. Subjective ATT Testing via Mindfulness Training

Domain: Consciousness Studies / Psychophysiology

Goal: Map how trained awareness affects ATT sensitivity.

Setup:

- Meditators vs. non-meditators perform the same perceptual or cognitive timing tasks.
- Measure variance in recognition time, clarity rating, and memory retention.

- Evaluate how silence, breathwork, and intention accelerate RAC formation.

Outcome:

- Confirm if internal stillness improves field alignment rate.

Insight: Waiting isn't inactivity. It's active preparation for clarity.

Summary Table: Future ATT Validation Experiments

Experiment Type	Domain	Core Mechanism	What It Tests
Cognitive Resonance Trials	Psychology	ATT under perception load	Recognition delay vs. visibility
Quantum Perception Match	Quantum Cognition	Delay between collapse + awareness	Conscious field effect on outcomes
EEG-Based ATT Mapping	Neuroscience	Brainwave response delay	ATT signature in neural timing
Phase-Matched ML Simulation	AI & Signal Systems	Filtered learning delay	Computational ATT model
Field Lag in Real Decisions	Behavioral Science	Decision vs. clarity lag	Ripple-induced decision error
Cosmological ATT Observations	Astrophysics	Signal delay via ripple logic	Large-scale perceptual timing
Mindfulness-Enhanced ATT Trials	Consciousness Studies	Awareness field expansion	Attentional control and ATT sensitivity

Final Note:

Validating ATT doesn't mean proving time passes. It means proving that clarity waits. These experiments aren't just scientific—they're existential. They remind us that life doesn't unfold in seconds. It unfolds in ripples—and we must learn to read the wave, not the clock.

"Before we observe reality, we must become its fit. Not everything waits to be seen—but everything waits to be aligned."

Glossary of Ripple Terms

A *Common Vocabulary for the Language of Alignment*

To navigate the universe through the lens of Universal Energy Dynamics (UED), one must understand its ripple-centric terminology. This glossary decodes the core concepts used throughout the book, clarifying the shift from conventional mechanics to field-based consequence dynamics.

Ripple

The fundamental mode of existence in UED. A ripple is a disturbance in the Structured Vacuum Energy (SVE) that carries consequence—not just energy. It represents an intent in motion seeking alignment.

SVE (Structured Vacuum Energy)

The still, intelligent background of the universe. It is not empty, but structured to allow ripples, alignment, and consequence. All actions, matter, and perception emerge from distortions in SVE.

ATT (Alignment Threshold Time)

The minimum duration needed for any ripple (stimulus, emotion, signal) to stabilize enough to be perceived by an observer. It's not reaction time, but resonance readiness.

RAC (Ripple Alignment Criterion)

The moment when a ripple fits perfectly with the observer's internal field. RAC is what allows clarity, recognition, and understanding. No RAC, no perception.

ΔΦ (Delta Phi – Phase Difference)

The mismatch in timing or rhythm between two interacting ripples (e.g., observer vs. field). A high $\Delta\Phi$ causes confusion, noise, or emotional friction.

$\Delta\tau$ (Delta Tau – Time Delay)

Temporal gap between ripple impact and observer perception. High $\Delta\tau$ means slow or failed realization—often felt as frustration, confusion, or surprise.

Ao / Ar (Amplitude Ratio)

Ao = Observer amplitude, Ar = Ripple amplitude. Their ratio reflects perception strength—how tuned in the observer is to the incoming ripple. A mismatch leads to poor clarity or emotional misreading.

Field Fit

The alignment condition between an observer's internal ripple field and an external signal or event. When field fit is achieved, RAC stabilizes and understanding occurs.

Ripple Misalignment

Occurs when incoming ripple data (sensory, emotional, cognitive) fails to meet ATT or RAC conditions. Results in confusion, stress, emotional pain, or error in action.

Ripple Trust

The state of confidence between ripples—especially the observer's trust in internal and external field stability. It governs intuition, presence, and insight.

Perception Window

A temporary opening in consciousness when ATT is met and RAC forms. This is when clarity and meaningful experience arise.

Ripple Collapse

Not to be confused with wavefunction collapse. Ripple collapse refers to the dissipation or absorption of a ripple's potential after alignment is denied or after consequence is released.

Consequence Vector

The direction and impact of an unfolding ripple event in space-time. Unlike force, it carries intent and meaning, not just motion.

Ripple Echo

A returning or lingering misalignment, often experienced as emotional residue (guilt, trauma, nostalgia). It's not memory—it's an uncompleted ripple.

Ripple Observer

An entity (human, AI, consciousness) whose perception and action depend on alignment with ripple fields. All sentient awareness is ripple-mediated in UED.

Cosmic ATT

The macro-scale alignment delay between universal intent and observable reality. Often mistaken for randomness or divine timing.

Ripple Permission

The field's natural gatekeeping mechanism. It decides when an event is allowed to be perceived, not when it happens.

Decoherence (UED Sense)

A loss of ripple alignment—not quantum superposition collapse. In UED, decoherence means perceptual noise, not probability resolution.

Phase Lock

A stabilized state where ripple frequencies sync—basis of deep attention, empathy, or insight.

Ripple Readiness

The pre-condition for RAC to form. It involves observer intent, emotional neutrality, clarity, and temporal availability.

Ripple Ladder

Refers to layered perception levels—from subconscious sensing to full meta-awareness. Each rung has a higher ATT and deeper RAC requirement.

Uncertainty (UED Sense)

Not probability, but field unreadiness. Uncertainty arises when RAC cannot stabilize due to $\Delta\Phi$, $\Delta\tau$, or amplitude mismatch.

Acknowledgments

For every ripple that reached me, and every silence that shaped me.

This book was not written alone.

It is a **consequence of many invisible alignments**—people, questions, tensions, encouragements, and quiet pauses that arrived just when they were meant to. I did not create this work from scratch; I simply tuned myself until it made sense. The ideas were already rippling in the universe—**I merely reached coherence.**

To my **readers and fellow observers**, thank you for trusting this unconventional path. In a world addicted to certainty, you chose to explore alignment. You've proven that curiosity is not dead—and that silence, too, has an audience.

To the **scientists and philosophers of the past**, whose brilliance shaped the frameworks I now respectfully challenge: thank you. Without your constructs, I would have had nothing to ripple against. You taught me that understanding begins not in agreement, but in questioning.

To my **peers and critics**—both real and digital—thank you for the friction. Every disagreement was a necessary phase shift. You reminded me that clarity is earned, not given.

To my **family**, who bore the strange rhythms of a mind often lost in ripple space: your patience is the field I trusted most. In your presence, I learned what it means to wait—not for approval, but for understanding.

To those who helped edit, question, refine, and challenge this manuscript—you know who you are—your alignment was critical.

Thank you for stepping into this field with me and staying long enough for coherence to form.

And to the **universe itself**—the stillness behind it all—thank you for letting me listen.

In this book, I did not try to explain everything. I only tried to **align with what already wanted to be said**. For those who resonated with it, you didn't just read—you rippled.

With deep gratitude,

Sandeep

About the Author

Sandeep Chavan is an independent researcher, educator, and prolific author whose work challenges the foundations of mainstream science, offering a powerful new framework—**Universal Energy Dynamics (UED)**—to decode reality through ripple alignment, not probability or force.

With over a hundred books across scientific, philosophical, and psychological domains, Sandeep is known for breaking through the walls of academic convention to present ideas that resonate with both logic and lived experience. His writings bridge ancient wisdom and modern inquiry, merging metaphysics with measurable consequence.

He has never been a passive follower of dominant narratives. Trained as an engineer and evolved as a thinker through collapse, teaching, and fierce introspection, he proposes a radical yet intuitive view of the universe—**a still field of structured potential, where consequence emerges through readiness, not randomness.**

Sandeep's signature style blends deep insight with conversational clarity. He's not afraid to ask the uncomfortable questions or rewrite the assumed laws of physics if they no longer serve. Whether exploring the ripple behind emotion, the silence behind action, or the true meaning of observation, he brings science back into alignment with human perception.

When he's not writing or teaching, Sandeep can often be found discussing reality under a tree, debating existence with students, or simply watching the world ripple—quietly, carefully, and curiously.

Don't miss out!

Visit the website below and you can sign up to receive emails whenever Sandeep Chavan publishes a new book. There's no charge and no obligation.

https://books2read.com/r/B-A-EPGPC-JNMQG

BOOKS 2 READ

Connecting independent readers to independent writers.

Did you love *Chavan–Heisenberg Ripple Uncertainty Principle*? Then you should read *Quantum Courtroom: The Trial of a Century*[1] by Sandeep Chavan!

[2]

What if everything we've been told about quantum mechanics—collapse, uncertainty, entanglement—is built on a misreading of perception?

Quantum Courtroom is not just a book. It's a trial. A relentless, cross-examining takedown of conventional quantum theory, where equations stand accused, scientists testify posthumously, and the observer finally takes the stand.

At the heart of this disruptive work lies the **Universal Energy Dynamics (UED)** framework, a consequence-first philosophy where

1. https://books2read.com/u/br8qwe

2. https://books2read.com/u/br8qwe

reality doesn't flow—it aligns. Waves don't travel—they ripple. Time doesn't pass—it triggers. And physics? It doesn't dictate—it emerges.

From Schrödinger's misunderstood equation to the myth of Born's probability, from Heisenberg's uncertainty to Bohr's epistemological confusion—every sacred cow is respectfully but rigorously challenged. In this courtroom, perception is not passive. Measurement is not neutral. Constants are not divine. And particles? They never existed. Only consequences did.

In five compelling parts, *Quantum Courtroom* leads the reader through:

The Case Against Collapse – exposing the duality myth and measurement illusion**The Collapse of Collapse** – unraveling uncertainty, entanglement, and the Copenhagen blind spot**The Ripple Ruling** – presenting structured vacuum and ripple logic as foundational truths**Rewriting Reality** – replacing force, constants, and time with alignment-based understanding**Reshaping Science** – offering a vision for future education, technology, and ethics

This is a bold manifesto for those who have sensed something's missing in modern physics—a deeper logic behind randomness, a role for intelligence in nature, and a call to restore meaning to observation itself.

Written in clear, engaging language—with no heavy math, just sharp insight—it's a must-read for scientists, philosophers, students, and independent thinkers alike.

Welcome to the courtroom where collapse collapses.And consequence takes the bench.

Read more at https://www.sandeepjchavan.com/.

Also by Sandeep Chavan

Chavan's UED research

CMGED Research Report

Chavan's Cosmic Continuum (C^3)

Frequency

Ripples of Reality

Reality Realigned

The Great Realignment

I – The Cosmic Leverage

Chavan's Cosmic Theory of Relativity (CTR)

Chavan's Theory of Light

Feeling Like a Physicist

TIME – The New Definition

$E \neq mc^2$: The Collapse of Classical Assumptions

The Chavan Codex

Chavan's Atomic Model

Echoes of War: The Russia-Ukraine Conflict and Its Global Ripples

Shattered Lives

The War Economy

Borders Redrawn

Arms and Allies

Collateral Damage
Justice in the Ashes

Somewhere in the Darkness of the Cosmos
The Blueprint of the Universe
The Cosmic Breath
From the Edge of the Universe

The Eternal Sage Chronicles
The Sage and the Sovereign
The Family Pillar
The People's Flame
The Warrior's Soul
The Bridging Mentor
The Guiding Light

The Exodus
The Last Cycle: 2075

The Fundamental Theory of Life & Existence
Energy in Action: The Soul Demystified
The Active Mind
Energy in Connection
Death: The Disconnection
Subconsciousness: The Chosen Disconnection
Desires: The Engine of Life

The Restoration of Human Knowledge

Restoring Bhagavad Gita

The War of Thoughts

Cracks in the Known

The Human Game

The Divine Dilemma

The Original Intent

The Answers

Ultra Modern Life Philosophy

The Duality of Laughter and Tears

Take It Easy

Compare to Accept

Standalone

It's Not AI, It's AHI - Amplified Human Intelligence

The Decision Paradox

The Four Sapiens

Malicious Script of Indian Polity

The IIT Legacy & Global Impact

The IIT Legacy & Global Impact

You are just a Version of Your Original

Win the Game You Didn't Choose

Always Keep Your Bags Packed

Reveal to Shield - Learn the Game of Facades

Engage Beyond Elections
Culture, Identity & Change: The Evolution of Indian Society
The Triangular Dynamics
Never Trust Without Doubt
If You Can't Change the Country, Leave the Country
Live Fearless with Fear
India: The Investment Magnet
Realignment
Powering a Billion Dreams
Universal Skill
Extension: From Brush to Algorithm
Human Side of Einstein
The Future Knows No Lines
Let's Bring the Generations to the Table
Unified Energy Dynamics
A Beat That Never Dies
Strategic Horizons
Right From Scratch
The Forgotten Innovators
Physics Fixing
The Savior Illusion
The Selfish Side
The Age of the Weakest Sapiens
Neural Wars
The Absorption: There Is No Death
Strange Therapist
Bubbles: Become a Drop
Beyond the Blue
Half-Scholar
The Law of Truth
Failure
Philosophy 2.0: A Structural Manifesto
Chaos: A Gentle Guide to the Unresolvable

Self-Help: New Syllabus
Rippling
Game: How to Play the Share Market
Out of Syllabus
Saffronized: The Code Behind the Crown
The Anger
Moksha: The Liberation
Chavan vs Einstein: The Universe Was Never Curved
Rocket Science: The Ripple Gambling
The Cosmos: The Still Universe
The Taste of Silence
Chavan's Post-Materialist Human Model
10-Second Rule
Chavan vs Hawking
It was Never a Wave Signal
Quantum Courtroom: The Trial of a Century
Chavan–Heisenberg Ripple Uncertainty Principle

Watch for more at https://www.sandeepjchavan.com/.

www.ingramcontent.com/pod-product-compliance
Lightning Source LLC
LaVergne TN
LVHW090320210926
848480LV00001B/308